Alloy Steel - Properties and Use

Alloy Steel - Properties and Use

Editor

Leroy Sidney

scitus
academics

Alloy Steel - Properties and Use
Edited by **Leroy Sidney**

ISBN: 978-1-68117-311-5
Library of Congress Control Number: 2016937620

SCITUS Academics LLC,
www.scitusacademics.com
Box No. 4766, 616 Corporate Way,
Suite 2, Valley Cottage,
NY 10989

Notice

Printed in United States of America on Acid Free Paper

Preface

Steel alloy is alloyed in a variety of elements between 1.0 and 50% by weight in order to improve its mechanical properties. There are two groups of alloy steels: low and high alloy steels. The term "alloy steel" refers mostly to low alloy steels. Alloy steels are made of carbon steel in combination with one or more alloy elements, including manganese, silicone, nickel, titanium, copper, chromium and aluminum. These metals are added in order to producc specific features not found in normal carbon steel. The elements are added in different proportions (or combinations) so that the material takes on various aspects such as higher hardness, higher corrosion resistance, higher strength and improved formability. To achieve some material properties, alloy elements are added. As a guideline, alloy elements are added in lower percentages to increase strength or durability or in higher percentages to achieve special properties such as resistance to corrosion or extreme temperature. During the steel production process, manganese, silicone or aluminum are added to remove dissolved oxygen, sulfur and phosphorus from the melt. Manganese, silicone, nickel and copper are added by the formation of solid ferrite solutions to increase strength. In the second phase, chromium, vanadium, molybdenum and tungsten increase their strength by forming carbides. Copper and nickel improve corrosion resistance in small amounts. Molybdenum helps to withstand vulnerability. Zirconium, cerium and calcium by controlling the shape increase the toughness of the inclusions. Sulfur increases machinability in manganese sulfide, plum, bismuth, selenium and tellurium form. Alloy elements also affect the temperature of the steel eutectoid. This book focuses on new approaches and uses of stainless steels, as well as new insights into strengthening mechanisms and toughness in micro-alloyed steels.

Table of Contents

CHAPTER 1

Effect of Heat Treatment Parameters on the Mechanical and Microstructure Properties of Low-Alloy Steel

Mohamed H. Frihat

Mechanical Engineering Department, Al-Hsoun University College, Al-Balqa Applied University, Irbid, Jordan

ABSTRACT

This study examines the effect of heat treatment at three different temperatures of 800°C, 950°C and 1100°C on the microstructure and mechanical properties of low-alloy steel with an addition of manganese, chrome and lead. To determine an impact of the applied heat treatment operations, testing of mechanical properties and microstructural examinations of the steel with 0.23%, 0.24%, 0.29% and 0.31% C were conducted. This work shows that the mechanical strengths of the alloy steel are improved with increasing the heat treatment temperature. In addition, the microstructure trends toward recrystallized ferrite grains as the heat treatment temperature increases.

KEYWORDS

Low-Alloy Steel, Heat Treatment, Microstructure, Mechanical Properties

1. INTRODUCTION

During the last decays, there has been a great demand for steels with higher mechanical strength, sufficient ductility and toughness. Moreover, the lightness of the steel is attractive, as in the automobile and aircraft applications. These requirements can be achieved by an increase in carbon content in a limited way, but even in the heat-treated condition the maximum strength of alloy steel can reach 700 MPa above this value; the ductility dramatically decreases [1] .

The heat treatment of alloy steels provides a high strength and yield point, combined with significant ductility even in large sections. Only thin section of plain carbon steels can be hardened via water quenching, and this is accompanied by a distortion and cracking. Alloy steels can resist corrosion and oxidation at high temperatures.

According to the Alloy Steels Research Committee (ASRC): "Carbon steels are regarded as steels containing not more than 0.5% manganese and 0.5% silicon, all other steels being regarded as alloy steels [2] ". The basic alloying elements added to steel are manganese, lead, nickel, chromium, molybdenum, vanadium, niobium, silicon and cobalt. The effect heat treatment of low-alloy steel was extensively studied by many researchers during the last decays. In the next paragraphs, samples of these studies are presented.

R. Roberti [3] carried out a research work to measure fracture toughness of ASME SA-542/SA-542M steel subjected to different tempering conditions to assess the effect of treatment temperatures higher than the maximum level for preweld heat treatment (PWHT) on both microstructure and toughness. It was found that, only small modifications in the microstructure are without any damage in the fracture behavior of base metal, weld metal.

Mohammad A. [4] investigated the effects of Cr and Ni on low carbon steel with undissolved carbide particles on the refining the austenite grain size. In the presence of nickel, chromium carbide is less active in austenite grain refinement than chromium carbide in absence of nickel at temperature under 975°C. Nickel does not produce any austenite grain refinement, but the presence of nickel promotes the formation of acicular ferrites.

Petr et al. [5] considered a secondary hardening of steels containing vanadium with non-tempered welded joints and he found that it experiences a significant increase in hardness during exposure at operating temperature.

Trindade Filho et al. [6] studied four different low-alloy steel weld metals (WM) both in the welded condition and after normalizing heat treatment. The alloy steel weld metals' mechanical properties, tensile and Charpy V toughness testing and microhardness measurements were evaluated. His results show the effect of normalizing heat treatment on welds metal properties: the original as welded metal fine grained microstructure is changed to a coarse equiaxed ferrite with ferrite-carbide aggregates and the yield tensile strength properties are considerably reduced.

Kalandyk and Zapała [7] carried out a study to investigate the effect of prolonged time of holding at the temperature of 620°C on the processes of secondary phase precipitation and mechanical properties of low-alloy cast steel with an addition of vanadium subjected to two variants of heat treatment. This study shows that the application of complex heat treatment holding at the temperature of 620°C for 10 hours cannot guarantee the required improvement of mechanical properties compared with the common type of heat treatment during which the time of holding at a given temperature is adjusted to the casting wall thickness.

Pang [8] studied the form of the hardness gradient in the heat affected zones (HAZs) produced by submerged arc welding of two low-carbon Q & T steels. Their results were shown unequivocally that the gradient differs from that found in steels of lower carbon equivalent in that the peak HAZ hardness was displaced from the grain coarsened heat affected zone (GCHAZ) into the grain refined heat affected zone (GRHAZ).

Ebert and Winsor [9] showed background information on the subject of tensile strength and resistance to wet H_2S cracking of carbon steel submerged arc welds obtained from reviewing industrial alloy steel weld metals' mechanical properties, tensile and Charpy V toughness testing and microhardness measurements.

Hamedi M. [10] studied the effect of process parameters including heating and post-weld heating power, and their corresponding duration along with interference, on the tensile strength of the welded joint was experimentally investigated.

It was found that with increasing the current, mechanical strength of the joint will increase to a specific limit and then decrease with a mild gradient. The mechanical strength of the joint increases with the time of heating; this behavior will improve to a specific limit and then subside to a particular value. Increasing the current augments the mechanical strength of the joint to a specific limit preceding a mild decrease afterward. The effect of post-welding heat on the tensile strength is evaluated and shows that this parameter has a remarkable effect exceeding 50% improvement in tensile strength. Increase in the interface of abutting surface causes the mechanical strength of the joint to increase. The study shows the direct effect of time and current in heating stage on mechanical strength with a steeper gradient. Therefore it can be stated that these parameters have a strong effect on the UW process.

Rasool and Zaidi [11] studied the influence of post weld heat treatment on the fracture toughness of low-alloy steel weldments using standard COTD test and reported correlated results.

This study showed a higher fracture toughness due to material homogeneity, as welded weld metal has lower COTD values. These properties are in high priority for structural steel if these can be blended with high tensile strength. When comparing the base metal and HAZ, the base metal showed higher fracture toughness compared with welded HAZ.

In the present work four different steel alloys with varying weight percentages of Ni and Cr were tested to investigate the effects of alloying elements. The effects were characterized according to the presence of alloying elements either alone or in conjunction with each other in the low carbon steel samples. The steel samples were characterized using mechanical testing methods. Tensile strength and elongation data were collected to study the effects of heat treatment and alloying elements. Metallographic and microstructure analysis was also carried out to draw a correlation between alloying elements and microstructural features.

2. EXPERIMENTAL WORK

2.1. Samples Preparation

Four different steels containing about 0.11% carbon were used in this experimental study. Table 1shows the composition of the steels alloy. Steel 1 is the base steel with which the structure and properties of other steels 2, 3 and 4 are compared. 2 samples have diameter 10 mm and 2 samples have diameter 14 mm with deferent chemical composition. Thus each sample has specific mechanical and microstructure properties of each sample.

The steels were made previously in an air induction furnace. About 14 mm diameter specimens were rolled down from 16 mm diameter bars of each of the steels in order to study the austenite grain coarsening behavior.

2.2. Pre-Test Operation

As mentioned previously four rods of steel with deferent chemical compositions, the work was divided into four groups with one rod in each, this rod was cut in to four pieces.

For tensile test the work piece was of length 30 cm, and for impact test the work piece has a dimension (5.5 cm × 1 cm × 1 cm) these samples were made by CNC milling machine and a Charpy notch was made in the center of length and angel 45° and depth 2 mm by using a lath machine as shown in the Figure 1.

The microstructure work piece was cut along of its length 1.5 cm and a machine was used to make moulds of plastic as shown as in Figure 2, which helps to catch the specimen easily on the testing machine called Moulding.

Table 1. (a) Chemical analysis for specimen 1; (b) Show chemical analysis for specimen 2; (c) Show chemical analysis for specimen 3; (d) Show chemical analysis for specimen 4.

(a)

Diameter 10 mm	Chemical Analysis									Mechanical Properties
	Type of billet	C%	Mn%	Cr%	N%	P%	S%	Cu%	Ceq%	Standard
										Grade
					Max.	Max.	Max.			JS441
10	GR5SP	0.31	0.55	0.03	0.008	0.025	0.039	0.03	0.41	420

(b)

Diameter 10 mm	Chemical Analysis									Mechanical Properties
	Type of billet	C%	Mn%	Cr%	N%	P%	S%	Cu%	Ceq%	Standard
										Grade
					Max.	Max.	Max.			BS4449
10	GR(500)	0.24	0.66	0.13	0.006	0.020	0.050	0.41	0.41	500

(c)

Diameter mm	Chemical Analysis									Mechanical Properties
	Type of billet	C%	Mn%	Cr%	N%	P%	S%	Cu%	Ceq%	Standard
										Grade
					Max.	Max.	Max.			BS4449
14	GR(500)	0.23	0.53	0.13	0.009	0.020	0.025	0.35	0.38	500

(d)

Diameter mm	Chemical Analysis									Mechanical Properties
	Type of billet	C%	Mn%	Cr%	N%	P%	S%	Cu%	Ceq%	Standard
										Grade
					Max.	Max.	Max.			JS441
14	GR 5SP	0.29	0.57	0.05		0.021	0.020	0.23		300

Figure 1. Impact piece.

Figure 2. Moulding unit.

A pail of water was used to immersion the sample rod after heat treatment, this operation knows as quenching.

2.3. Carburization and Measurement of Austenite Grain Size

Since size of the austenite grains directly affect the subsequent structure and hence the properties of steels, a study was made to determine prior austenite grain size at temperatures higher than upper critical temperature Carburization technique was used to reveal prior austenite grain size. There are also other methods in determining prior austenite grain size like isothermal transformation technique, oxidation technique. But previous work showed that the isothermal technique did not work well in revealing prior austenite grain boundary of low alloy steels [12]. Therefore, carburization technique was adopted to reveal prior austenite grain boundaries of steels in this work.

The technique is based on the formation of a continuous cementite network at the austenite grain boundaries. Carbon will diffuse in steel from the carburizing atmosphere forming hypereutectoid steel at the surface of the specimen and during slow cooling in the furnace continuous cementite network is formed at the austenite grain boundaries at the selected austenitizing temperatures.

Subsequent etching of the furnace cooled samples revealed the cementite network formed which marked the prior austenite grain size at the selected carburizing temperatures [13] .

Solid carburizing or pack carburizing technique was applied in this experimental study. The steel specimens were heated to different austenitizing temperatures, i.e. 900˚C - 1050˚C with an interval of 50˚C. Before heating these specimens, they were packed in a pot with carburizing mixture. Then they were placed in Blue-M furnace.

After reaching the desired temperature, they were held at that temperature for 2 h to reach near the equilbrium conditions and then cooled in the furnace to the room temperature. Slow cooling ensured a continuous cementite network through the austenite grain boundaries.

The assessment of prior austenite grain size was made from direct measurement of the austenite grains in the specimens under optical microscope. The grain size was measured using the mean linear intercept method, counting grain boundary intersections with the circumference of the circle in the eyepiece of a microscope. The effective circumference of the circle was determined precisely by measuring its diameters with reference to a stage micrometer at the magnification used. A total of at least 300 - 600 intersections were counted for each specimen. Then the size of austenite grain was measured using the mean linear method.

3. EXPERIMENTAL RESULTS

Tensile testing is a fundamental materials test in which a sample is subjected to a controlled tension until failure. The results from the test are commonly used to select a material for an application, quality control, and to predict how a material will react under other types of forces. Properties that are directly measured through a tensile test are ultimate tensile strength, maximum elongation and reduction in area. From these measurements the following properties can be determined: modulus, Poisson's, yield strength, and strain-hardening characteristics. Uniaxial tensile testing is the most commonly used for obtaining the mechanical properties.

A tensile specimen as shown as in Figure 3 with a standardized sample cross-section. It has two shoulders and a gauge (section) in between. The shoulders are large so they can be readily gripped, whereas the gauge section has a smaller cross-section so that the deformation and failure can occur in it. Figure 4 illustrates the most common types of shoulders. Keys A through C are for round specimens, whereas keys D and E are for flat specimens.

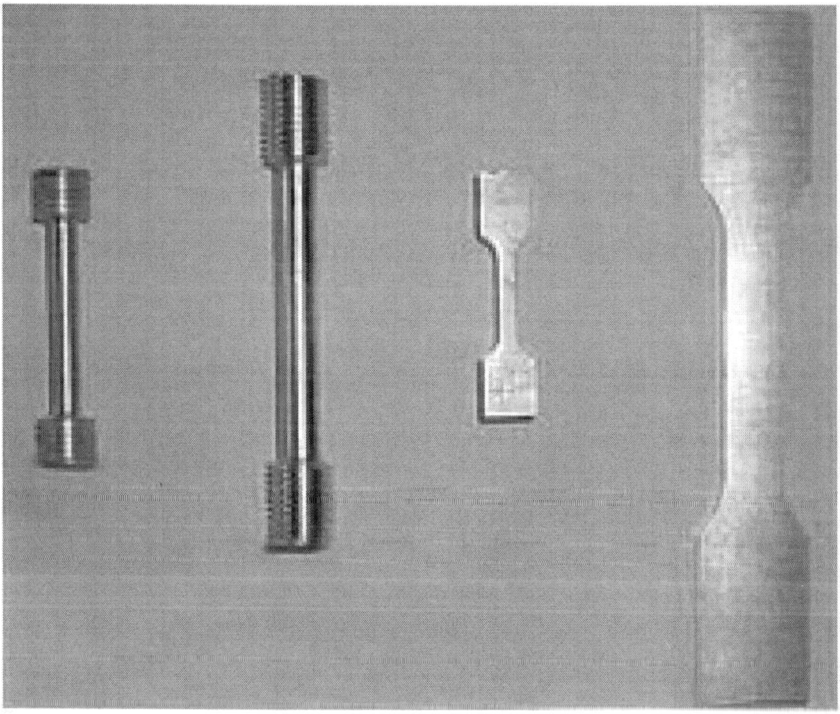

Figure 3. Moulding unit.

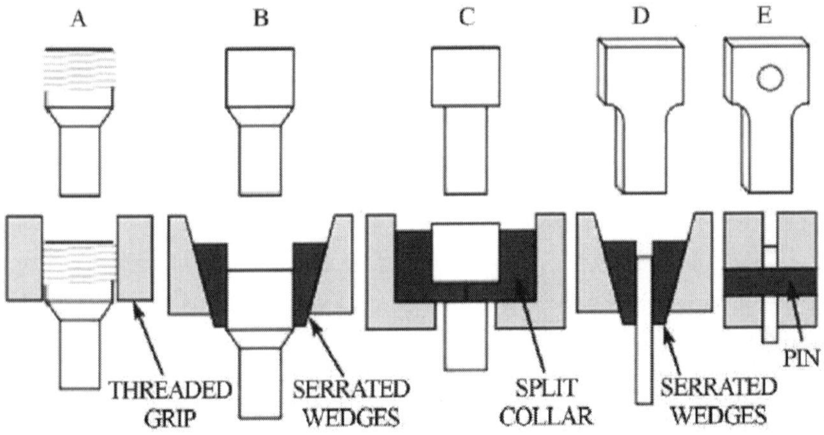

Figure 4. Various shoulder styles.

A standard specimen is prepared in a round section along the gauge length, depending on the standard used. Both ends of the specimens should have sufficient length and a surface condition such that they are firmly gripped during testing. The initial gauge length Lo is standardized (in several countries) and varies with the diameter (Do) or the cross-sectional area (Ao) of the specimen as shown as in Figure 5.

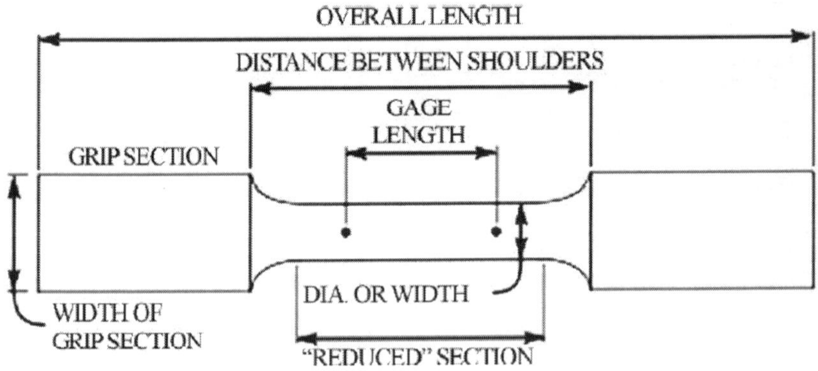

Figure 5. Test specimen nomenclature.

3.1. Testing Equipments

The most common testing machine used in tensile testing is the universal testing machine as shown as in Figure 6.

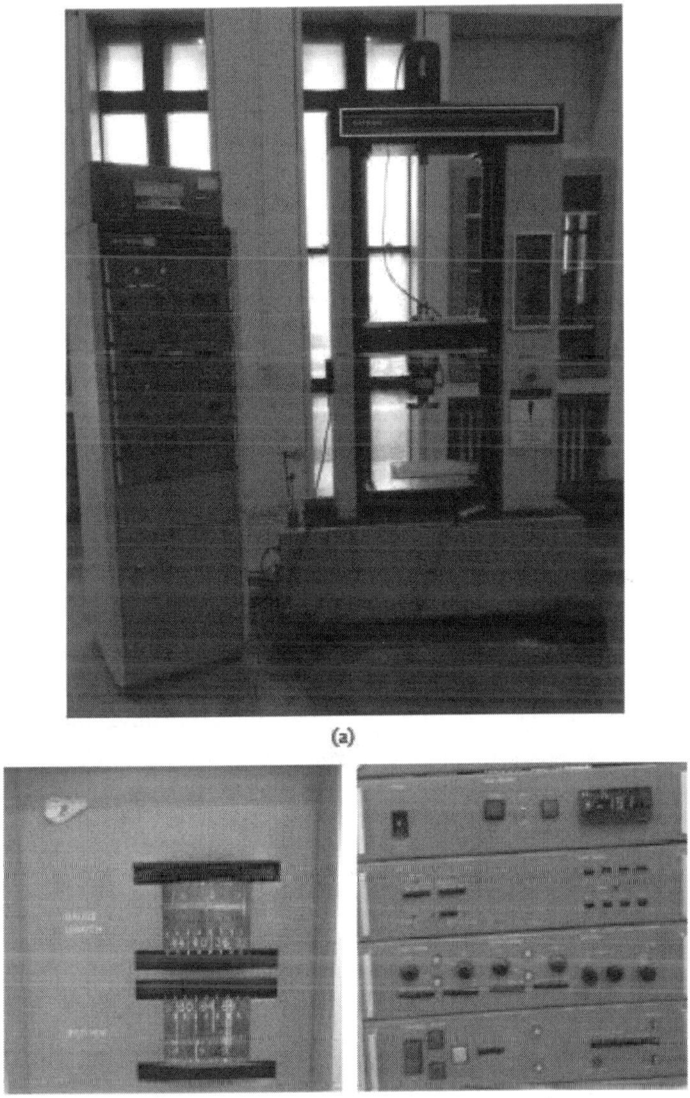

(a) (b) (c)

Figure 6. (a) A universal testing machine (Hegewald & Peschke); (b) Gauge length; (c) Panel load control.

Alignment of the test specimen in the testing machine is critical, because if the specimen is misaligned, either at an angle or offset to one side, the machine will exert a bending force on the specimen. This is especially bad for brittle materials, because it will dramatically skew the results. This situation can be minimized by using spherical seats or U-joints between the grips and the test machine. If the initial portion of the stress-strain curve is curved and not linear, it indicates the specimen is misaligned in the testing machine.

The strain measurements are most commonly measured with an extensometer, but strain gauges are also frequently used on small test specimen or when Poisson's ratio is being measured. Newer test machines have digital time, force, and elongation measurement systems consisting of electronic sensors connected to a data collection device (often a computer) and software to manipulate and output the data. However, analog machines continue to meet and exceed ASTM, NIST, and ASM metal tensile testing accuracy requirements, continuing to be used today.

3.1.1. Stress-Strain Curve

The relationship between the stress and strain that a particular material displays is known as that particular material's stress-strain curve. It is unique for each material and is found by recording the amount of deformation (strain) at distinct intervals of tensile or compressive loading.

Stress-strain curves of various materials vary widely, and different tensile tests conducted on the same material yield different results, depending upon the temperature of the specimen and the speed of the loading. It is possible, however, to distinguish some common characteristics among the stress-strain curves of various groups of materials and, on this basis, to divide materials into two broad categories; namely, the ductile materials and the brittle materials.

Figure 7 shows a typical stress-strain curve showing yield behavior for nonferrous alloys. Stress (σ) is shown as a function of strain ($\in$). Point

1: Elastic (proportionality) limits and point 2: Offset yield strength (0.2% proof strength.

3.1.2. Impact Test

Impact testing is testing an object's ability to resist high-rate loading. An impact test is a test for determining the energy absorbed in fracturing a test piece at high speed. Impact resistance is one of the most important properties for a part designer to consider, and without question, the most difficult to quantify. The impact resistance of a part is, in many applications, a critical measure of service life. More importantly these days, it involves the perplexing problem of product safety and liability.

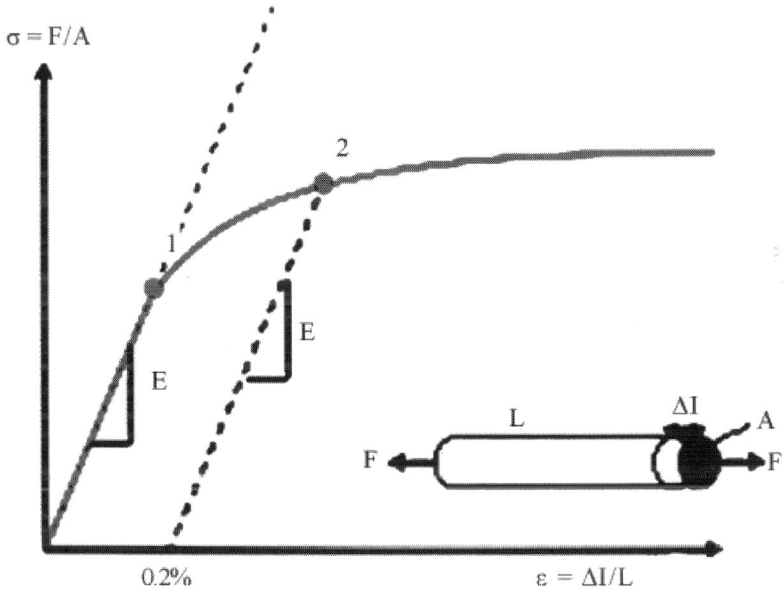

Figure 7. Stress-strain curve showing typical yield behavior for nonferrous alloys.

The Charpy impact testis a standardized high strain-rate test which determines the amount of energy absorbed by a material during fracture. This absorbed energy is a measure of a given material's notch toughness and acts as a tool to study temperature-dependent ductile-brittle transition. It is widely applied in industry, since it is easy to prepare and conduct with a results can be obtained quickly and cheaply. A disadvantage is that some results are only comparative, the impact test machine as shown as in Figure 8.

Figure 8. A vintage impact test machine.

3.1.3. Microstructure

Microstructure is defined as the structure of a prepared surface or thin foil of material as revealed by a microscope as shown as in Figure

9 above 25× magnification. The microstructure of a material can strongly influence physical properties such as strength, toughness, ductility, hardness, corrosion resistance, high/low temperature behavior, wear resistance, and so on, which in turn govern the application of these materials in industrial practice. Microstructure at scales smaller than can be viewed with optical microscopes is often called ultra structure or nanostructure.

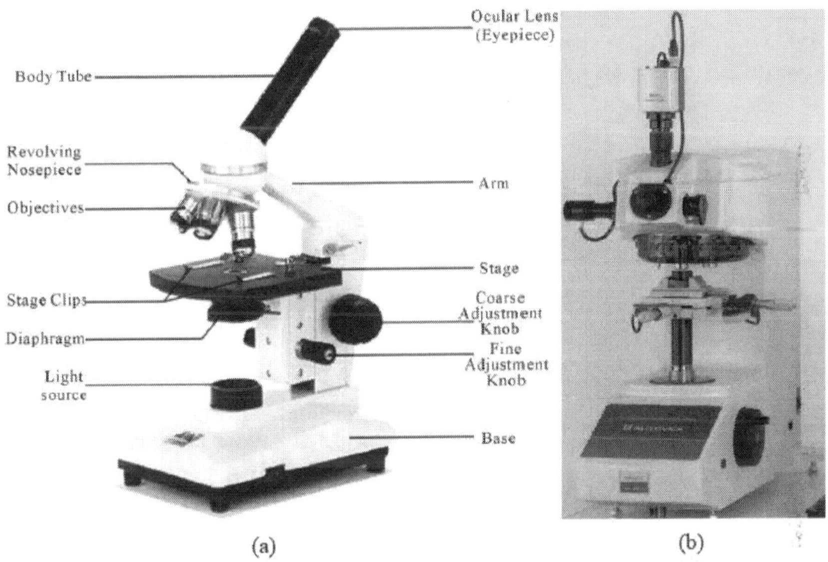

Figure 9. (a) Microscope & parts; (b) Micro Vickers hardness testing machines.

After preparing the specimens (cutting and make a moulds) we are grinding it by using paper grinding which have deferent degrees (60 - 120 - 180 - 220 - 320 - 420) where the increase in the degrees indicate the increase smoothness, after that we are polishing it using the paper polisher have great degree (1200) with diamond particles, then putting in nitric acid to indicate the grain boundaries. After every step were the specimens washed with alcohol. After that we use a microscope to see the grains and grain boundaries.

3.1.4. Microvickers Hardness

The Vickers hardness test method, also referred to as a micro hardness test method, is mostly used for small parts, thin sections, or case depth work. The Vickers method is based on an optical measurement system. The Micro hardness test procedure specifies a range of light loads using a diamond indenter to make an indentation which is measured and converted to a hardness value. It is very useful for testing on a wide type of materials as long as test samples are carefully prepared. A square base pyramid shaped diamond is used for testing in the Vickers scale. Typically loads are very light, ranging from a few grams to one or several kilograms, although “Macro” Vickers loads can range up to 30 kg or more. The Micro hardness methods are used to test on metals, ceramics, and composites—almost any type of material.

Since the test indentation is very small in a Vickers test, it is useful for a variety of applications: testing very thin materials like foils or measuring the surface of a part, small parts or small areas, measuring individual microstructures, or measuring the depth of case hardening by sectioning a part and making a series of indentations to describe a profile of the change in hardness. The Vickers method is more commonly used.

Sample preparation is usually necessary with a micro hardness test in order to provide a small enough specimen that can fit into the tester. Additionally, the sample preparation will need to make the specimen’s surface smooth to permit a regular indentation shape and good measurement, and to ensure the sample can be held perpendicular to the indenter. Usually the prepared samples are mounted in a plastic medium to facilitate the preparation and testing. The indentations should be as large as possible to maximize the measurement resolution (error is magnified as indentation sizes decrease). The test procedure is subject to problems of operator influence on the test results, in Figure 9 show Micro Vickers Hardness Testing Machines.

4. RESULT AND DISCUSSION

4.1. The Result and Discussion of Tensile Test

Result and Discussion Sample 1

In the sample 1 we were test mechanical properties before Heat treatment and after heat treatment at (800˚C, 950˚C, 1100˚C), the result after test:

It was shown that the steel was ductile material before heat treatment as shows in Figure 10 the carve started as a liner line after that show nicking zone it was guided in upper yield point at (405.97 N/mm^2) and lower yield point at (405.35 N/mm^2) and after that fracture point at which the value of elongation equal 18.46%.

Figure 11 illustrate the stress strain curve after heat treatment, it is clear that the material became brittle as a raped cooling after quenching operation, and the value of elongation at 800˚C equal 0.94% and at 950˚C equal 0.83% and at 1100˚C equal 0.96% .

In the sample 2 were tested to obtain the mechanical properties before and after heat treatment at (800˚C, 950˚C, 1100˚C.

Figure 12 shows the stress-strain diagram of sample 2 before heat treatment. It can be seen that, the steel was ductile material, the carve started as liner line after that it shows nicking zone guided in upper yield point at (515.99 N/mm^2) and lower yield point at (507.47 N/mm^2). The value of elongation is equal 12.03% after the fracture point.

The stress strain curve for sample 2 after heat treatment is shown in Figure 13. It is obvious that the material became brittle due to raped cooling or quenching operation. The elongation values decreases as the temperature increases. The value of elongation at 800˚C equal 1.25% and at 950˚C equal 0.94% and at 1100˚C equal 0.96%.

The stress strain diagrams of sample 3 before and after heat treatment are shown in Figure 14 andFigure 15 respectively. It can be seen that the curve started in liner line and then tend to nicking guided with an upper yield point at (566 N/mm^2) and lower yield point at (564.65/mm^2) finally the fracture point elongation equal is equal to 11.68%.

After heat treatment the material became brittle due to raped cooling and the value of elongation at 800°C is equal 1.02% and at 950°C is equal 1.15% and at 1100°C is equal 1.08% (see Figure 15).

Result Data	
Tensile Strength (Rm) N/mm^2:	676.04
Upper Yield (YeH) N/mm^2:	405.97
Lower Yield (YeL) N/mm^2:	405.35
Elongation% (A):	18.46

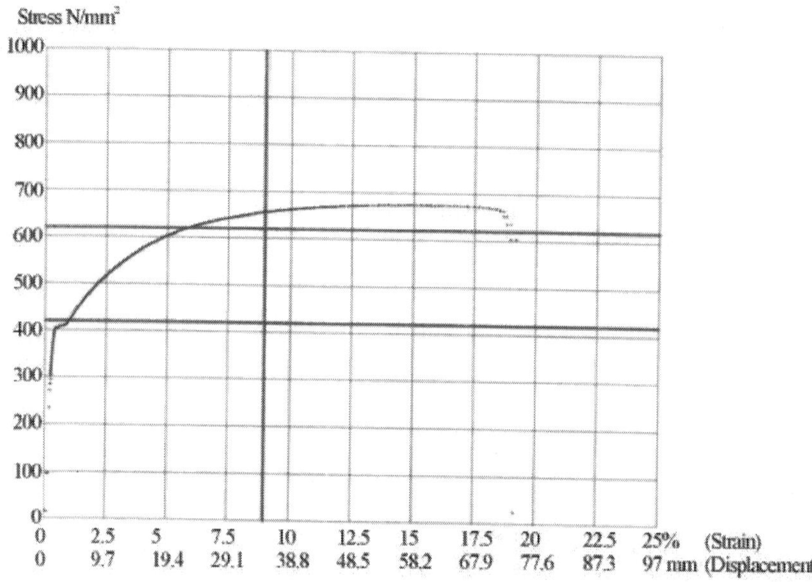

Figure 10. Sample 1 before heat treatment.

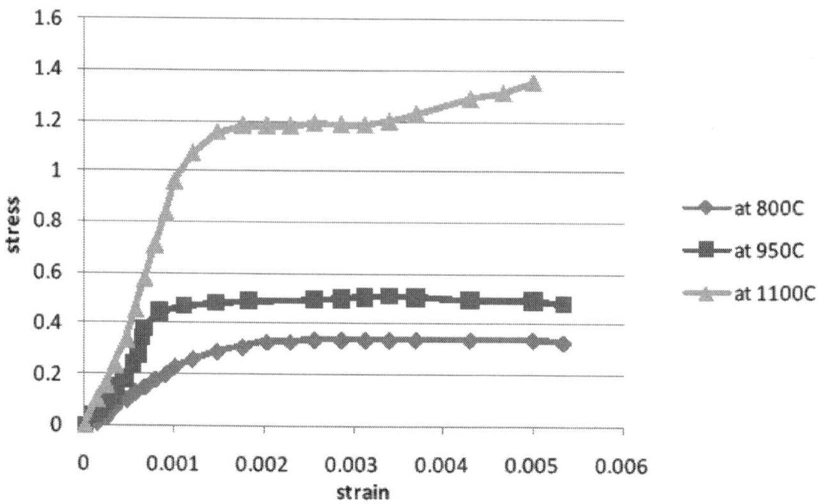

Figure 11. The result after heat treatment.

Result Data

Tensile Strength (Rm) N/mm²:	616.47
Upper Yield (YeH) N/mm²:	515.99
Lower Yield (YeL) N/mm²:	507.47
Elongation% (A):	12.03

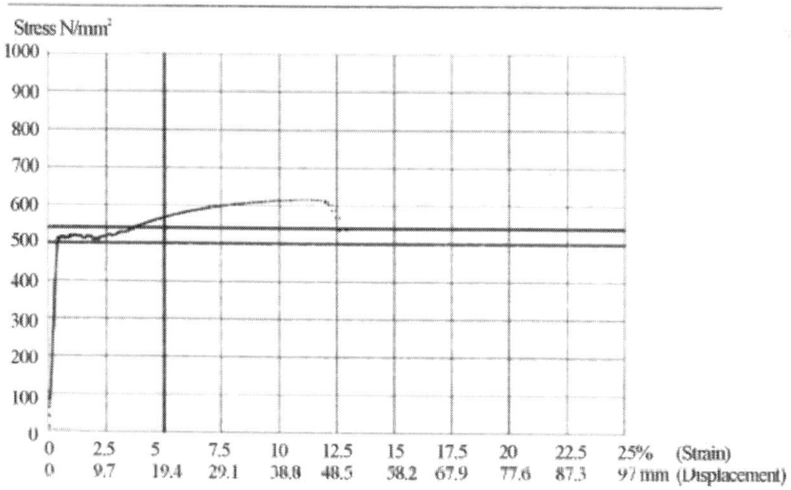

Figure 12. Sample 2 before heat treatment.

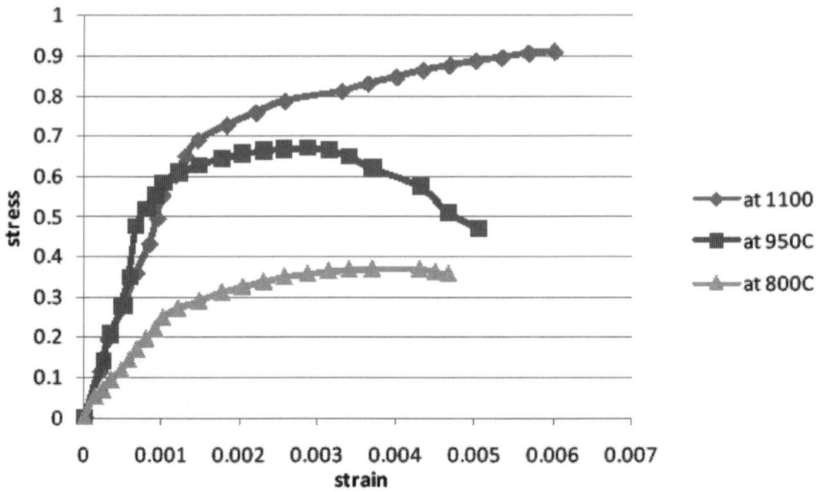

Figure 13. The result after heat treatment.

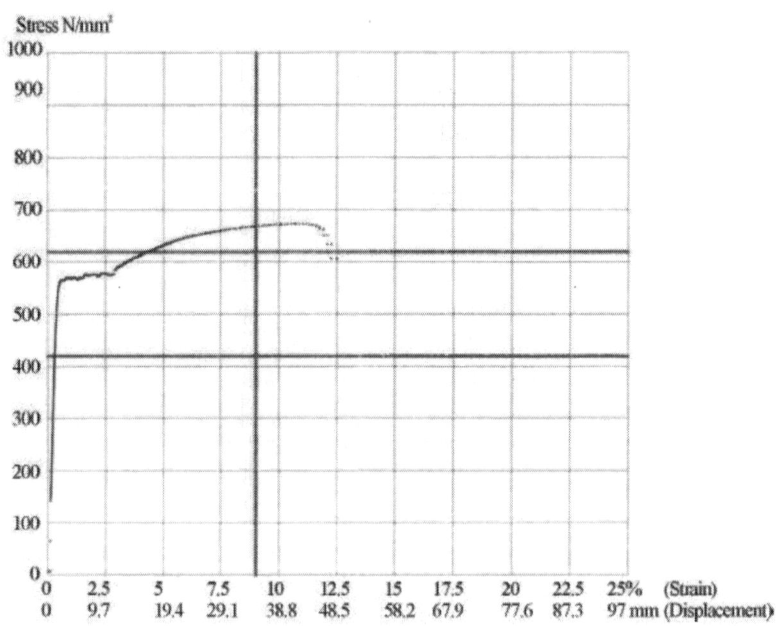

Figure 14. Sample 3 befor heat tretment.

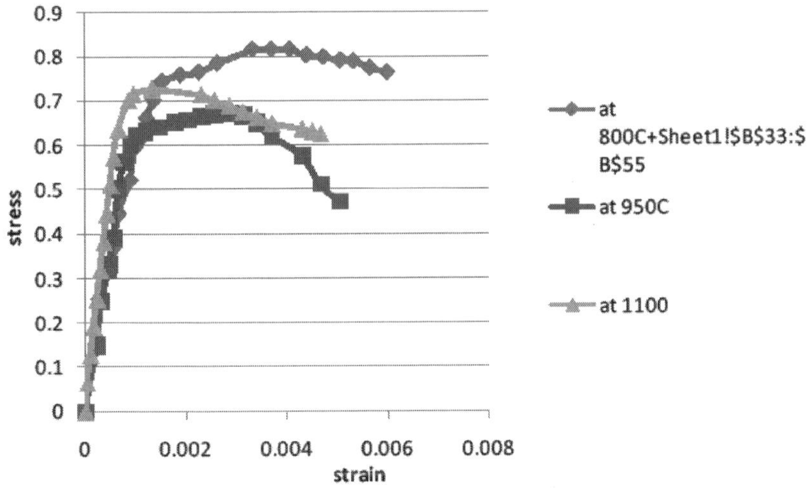

Figure 15. The result after heat treatment.

4.2. The Result and Discussion of Impact Test

Four samples of alloy steel were tested on the impact testing machine. The dimension of impact specimens: are: Charpy: $v_{effective} = 8 \times 40 \times 10 = 3200\ mm^3 = 3.2 \times 10^{-6}\ m^3$ (Figure 16).

Figure 17 illustrates the effect of heat treatment at different temperatures on the micro structure of sample 2.

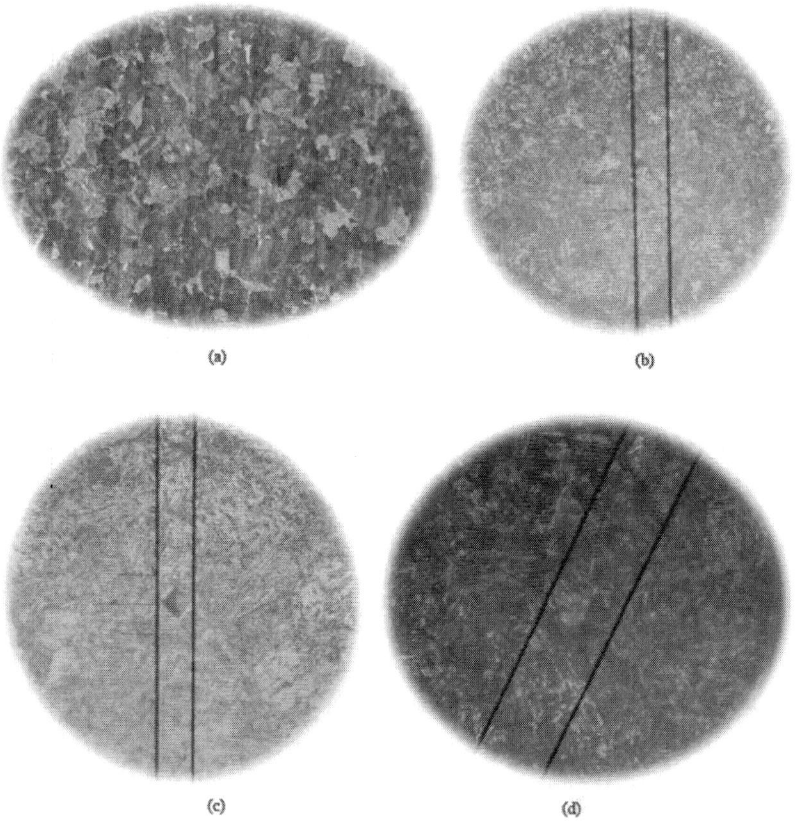

Figure 16. (a) Sample 1 microstructure before heat treatment; (b) Sample 1 microstructure after heat treatment at 800˚C; (c) Sample 1 microstructure after heat treatment at 950˚C; (d) Sample 1 microstructure after heat treatment at 1100˚C.

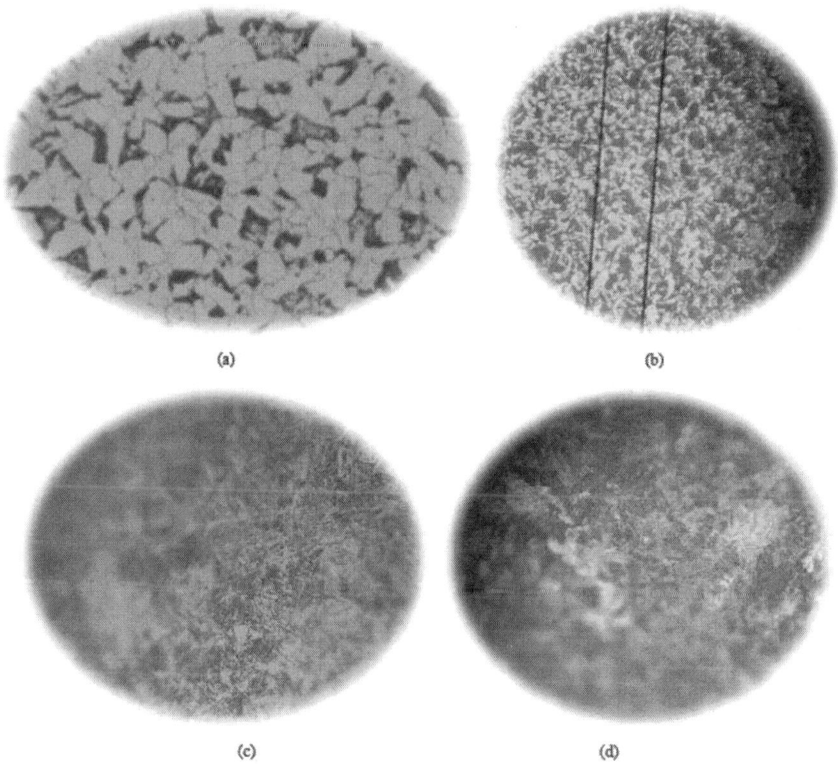

Figure 17. (a) Sample 3 microstructure before heat treatment; (b) Sample 3 microstructure after heat treatment at 800°C; (c) Sample 3 microstructure after heat treatment at 950°C; (d) Sample 3 microstructure after heat treatment at 1100°C.

5. CONCLUSION

The effect of heat treatment at three different temperatures of 800°C, 950°C and 1100°C on the microstructure and mechanical properties of low-alloy steel containing manganese, chrome and lead were experimentally investigated. In order to determine an impact of heat treatment operations, testing of mechanical properties and microstructural examinations of the steel with 0.23%, 0.24%, 0.29% and 0.31% C were carried out. This work shows that the mechanical strengths of the alloy steel are improved with increasing the heat treatment temperature. In addition, the microstructure trends toward recrystallized ferrite grains as the heat treatment temperature increases.

REFERENCES

1 Aver, S.H. (1974) Introduction to Physical Metallurgy. 2nd Edition, McGraw-Hill Book Company, New York.

2 American Society for Metals (1964) Metal Handbook Vol. 2. 8th Edition, American Society for Metals, Materials Park, Ohio.

3 Roberti, R., Cornacchia, G. and Faccoli, M. (2012) Effect of Increasing Post Weld Heat Treatment Temperature on the Fracture Toughness of an ASME SA-542M Steel. Forni di Sopra (UD), Italia.

4 Razzak, M.A. (2011) Heat Treatment and Effects of Cr and Ni in Low Alloy Steel. Bulletin of Materials Science, 34, 1439-1445.

5 Mohyla, P., Tomcík, P., Benes, L. and Hlavaty, I. (2011) Influence of Post Weld Heat Treatment on Secondary Hardening of CrMoV Welded Joints. Metal Science and Heat Treatment, 53, 374-378.

6 da Trindade Filho, V.B., Guimaraes, A.S., Payao Filho, J. da C. and Paranhos, R.P. da R. (2004) Normalizing Heat Treatment Effects on Low Alloy Steel Weld Metals. Journal of the Brazilian Society of Mechanical Sciences and Engineering, 26.

7 Kalandyk, B. and Zapala, R. (2008) Effect of Heat Treatment Parameters on the Properties of Low-Alloy Cast Steel with Microadditions of Vanadium. Archives of Foundry Engineering, 8, 137-140.

8 Pang, W., Ahmed, N. and Dunne, D. (2011) Hardness and Microstructural Gradients in the Heat Affected Zone of Welded Low-Carbon Quenched and Tempered Steels. Australasian Welding Journal, 56, 36-48.

9 Ebert, H.W. and Winsor, F. (1984) Carbon Steel Submerged Arc Welds-Tensile Strength vs. Corrosion Resistance. Welding Research Supplement 1193-s.

10 Hamedi, M. (2006) Optimizing Tensile Strength of Low-Alloy Steel Joints in Upset Welding. Journal of Achievements in Materials and Manufacturing Engineering, 17, 341-344.

11 Mohideen, R. and Ahmed Zaidi, A.M. (2010) Influence of Post Weld Heat Treatment on the HAZ of Low Alloy Steel Weldments. International Journal of Integrated (Issue on Mechanical, Material and Manufacturing Engineering).http://eprints. uthm.edu.my/ 616/1/ ijiev22010p2.pdf

12 Zahirul, H.M. (1989) Effect of Alloying Elements in HSLA. MsC Engg. Thesis, Department of Metallurgical Engineering, BUET, Dhaka, 130-150.

13 Clark, D.S. and Varney, W.R. (1962) Physical Metallurgy for Engineers. 2nd Edition, Litton Educational Publishing Inc., New York.

CHAPTER 2

Characterization of the Microstructure Evolution in IF-Steel and AA6016 during Plane-Strain Tension and Simple Shear

Gregory Gerstein [1], Benjamin Klusemann [2], Swantje Bargmann [2,3,*] and Mirko Schaper [4]

[1] Institute of Material Science, Leibniz Universität Hannover, Garbsen 30823, Germany

[2] Institute of Continuum Mechanics and Material Mechanics, Hamburg University of Technology, Hamburg 21073, Germany

[3] Institute of Materials Research, Helmholtz-Zentrum Geesthacht, Geesthacht 21502, Germany

[4] Lehrstuhl für Werkstoffkunde, Universität Paderborn, Paderborn 33098, Germany

ABSTRACT

In the current work, the evolutions of grain and dislocation microstructures are investigated on the basis of plane strain tension and simple shear tests for an interstitial free steel (DC06) and a 6000 series aluminum alloy (AA6016-T4). Both materials are commonly-used materials in the automobile industry. The focus of this contribution is on the characterization and comparison of the microstructure formation in DC06 and AA6016-T4. Our observations shed light on the active mechanisms at the micro scale governing the macroscopic response. This knowledge is of great importance to understand the physical deformation mechanisms, allowing the control and design of new, tailor-made materials with the desired material behavior.

KEYWORDS

microanalysis; mechanical characterization; sheet forming; fracture; electron microscopy

1. INTRODUCTION

In today's industrial manufacturing of sheet metal products, highly-specialized alloys are used to improve the properties of the final product and/or to reduce the total costs of production. The use of high strength steels and aluminum alloys is desired, because lower weight is achieved at comparable structural stiffness, e.g., in automotive body parts. In order to fully exploit these materials, deep drawing as a manufacturing process is simulated using continuum mechanical models and the finite element method. The material behavior is described by the constitutive formulation of the stress. Material models commonly used (e.g., based on the classical works of von Mises [1] and Hill [2] or more recent works by Barlat *et al.* [3]) describe the material behavior at the macroscopic scale in a phenomenological manner. Central to these models is that their physical motivation originates in the interpretation of the stress-strain data of mechanical tests, such as uniaxial tension or simple shear tests.

There exist several experimental investigations on the strain path-dependent behavior of interstitial free steel (IF steel); see, e.g., [4,5,6,7,8,9,10,11,12]. Ghosh and Backofen [6], for example, report an increase in yield stress for certain sequences of biaxial stretching. Later, these transitions in the stress-strain curves are related to the formation of persistent dislocation structures and their interactions with dislocations associated with the subsequent loading direction; see, e.g., [5,9,10]. Fernandes and Schmitt [5] show that under load path changes (uniaxial to biaxial tension) in low carbon steel, the dislocation structures are comparable to the ones under monotonic loading. Their work also indicates that the dislocation structure saturates for large strain values, as the morphology does not evolve with further applied strain. Wilson and Bate [11] find that the formed dislocation structure mainly determines the hardening behavior under load path changes; other properties, such as grain size, grain morphology and texture, are only of minor importance. The recent work of Yin *et al.* [12] investigates the material behavior of DC06 under shear conditions and provides information about the comparability of different shear test configurations and the homogeneity of the shear zone. van Riel [13] studied the mechanical behavior of DC06, dual-phase steel DP600, high strength steel H340LAD and aluminum alloy AA5182 under strain path changes. In addition to the observed Bauschinger effect (*i.e.*, a significant decrease of yield stress after reverse loading [14]), strong cross-hardening is observed for DC06 under orthogonal load path changes.

Aluminum alloy AA6016-T4 is investigated by Zillmann *et al.* [15]: the stress-strain behavior in uniaxial and biaxial compression tests is analyzed, and a biaxial tension-compression symmetry is observed. In AA6016, the Bauschinger effect is documented by shear tests with load reversal after different amounts of forward prestrain, *cf.* [16]. Hasegawa *et al.* [17] correlated the Bauschinger effect and the decrease in hardening rate after such a change with the evolution of the dislocation microstructure in pure aluminum. In particular, the decrease in hardening rate is linked to the partial dissolution of cells after load reversal.

Mathematical modeling approaches associated with such micro-macro interactions differ by the scale at which the mathematical constructs and their emerging relations are formulated. The microscale-based mechanistic approaches use mathematical entities directly linked to corresponding microstructural phenomena. In general, plastic yielding is related to the formation and interaction of dislocations in single-phase materials with few precipitates. As a consequence, a detailed model requires the modeling of the dislocation interaction on a scale of a few microns (e.g., [18,19,20,21,22,23,24,25,26,27]). However, in forming simulations, this is computationally too expensive. In phenomenological models, macroscopic work hardening behavior is expressed in terms of a function of equivalent plastic strain. The latter can be interpreted as a measure for the intensity of interactions of dislocations. Phenomenological models of this class yield acceptable results simulating sheet metal forming processes; see, e.g., Roll *et al.* [28] for single-phase materials, such as HC260LAD, Barthel *et al.* [29] for LH800 or Behrouzi *et al.* [30] for DC06. A recent model by Rabahallah *et al.* [31] includes crystallographic information on grain orientations in the computation of coefficients describing the initial point of yielding. However, most phenomenological models do not include crystallographic information in the description of plastic anisotropy resulting from texture.

Most of these models are not suited to relate the stress-strain curve for a given deformation path to the results gained by optical analysis of the grain and dislocation microstructure. Important contributions to the combination of knowledge gained from the investigations by transmission-electron microscopy (TEM) analysis of single and polycrystalline metals subjected to mechanical loading combined with a phenomenological modeling approach are made by the group of Teodosiu (e.g., Hu [32], Teodosiu and Hu [33], Bouvier *et al.* [34]). These works result in a phenomenological model for the transient behavior of polycrystalline sheet metals after changes in strain path.

Regardless of the chosen modeling approach, material modeling motivated by analysis of the microstructure requires the knowledge of active mechanisms at given levels of deformation and strain mode for different classes of materials. The goal of this investigation is to provide further insight into the microstructural evolution, *i.e.*, the dislocation patterns, in DC06 and AA6016-T4 for the future development of micromechanical motivated models for application in forming processes.

2. MATERIALS AND MATERIAL TESTING

2.1. Interstitial Free Steel DC06 and Aluminum Alloy AA6016

Interstitial free steel DC06 (sheet thickness 1 mm) delivered by ThyssenKrupp Steel Europe AG is investigated. Table 1states the chemical composition according to DIN EN 10130:2006 [35]. After cold rolling, the material was annealed and subjected to a final skin pass by the manufacturer. The average Young's modulus E over the three directions 0°, 45° and 90° with respect to the rolling direction was determined as 181 GPa with Poisson's ratio $v = 0.3$ [4].

Table 1. Chemical composition (mass %) of DC06 according to DIN EN 10130:2006 [35], determined by the manufacturer (ThyssenKrupp Steel Europe) by means of an analysis of one single specimen.

Alloying Component	C	Si	Mn	P	S	Al	N	Ti	Fe
weight percentage	0.003	0.018	0.137	0.013	0.010	0.035	0.0027	0.079	balance

The second material used in this investigation is an industrial aluminum alloy (AA6016 in T4 condition), in the form of rolled sheets of 1 mm thickness. The aluminum was obtained from Novelis (Sierre, Switzerland). It was annealed during manufacturing and not skin passed. The microstructure of the alloy in the as-received state is recrystallized. The composition of the material according to DIN EN 573-3:2007 [36] is given in Table 2. Young's modulus E is determined as 68 GPa with Poisson's ratio $v = 0.33$.

Table 2. Chemical composition (mass %) of AA6016 according to DIN EN 573-3:2007 [36].

Alloying Component	Si	Fe	Cu	Mn	Mg	Cr	Zn	Al
weight percentage	1–1.5	0.5	0.2	0.2	0.25–0.6	0.1	0.2	balance

2.2. Material Testing

Both materials were subjected to uniaxial tension, plane strain (ps) tension and simple shear loading. Simple shear tests were performed with the shear direction parallel and orthogonal to the rolling direction, where the rolling direction is parallel to the tension direction. The uniaxial tensile tests have been carried out at the Institute of Materials Science, Leibniz Universität Hannover, Germany. The remaining tests were performed with a biaxial testing device (see [37] (Figure 1) for a schematic image of the device) at the Institute of Engineering Technology at Twente University, Netherlands, which is capable of loading a sheet metal specimen in simple shear and in plane strain tension. The biaxial testing device consists of a uniaxial frame, as well as a sub-frame, which is mounted between the cross bars. The sub-frame accommodates an actuator for simple shear deformation. The possibility of buckling during simple shear was minimized by selecting an appropriate ratio of the height $b = 3$ mm in the deformation region to the sample thickness $t = 1$ mm. Homogeneous deformation was achieved in the measurement area due to a large ratio of width to height $c/b = 15$. The final shape of the specimen was obtained by wire eroding in order to minimize the influence of residual stresses at the edges. For further details of the experimental setup, the reader is referred to van Riel and van den Boogaard [37].

The tests were performed at constant nominal strain rates of 10^{-3} s^{-1}. The stress-strain responses for DC06 were previously reported in [4] (Figure 2) and the results for monotonic shear of AA6016-T4 in [38] (Figure 3); both showing a typical monotonic increasing stress-strain behavior. The initial yield strengths σ_Y (at $\varepsilon = 0.002$) are identified as $\sigma_{Y,\ DC06} = 145$ MPa and $\sigma_{Y,\ AA6016} = 117$ MPa. The ultimate tensile strengths σ_m are determined as $\sigma_{m,\ DC06} = 385$ MPa and $\sigma_{m,\ AA6016} = 239$ MPa. For completeness, the deformation states and the corresponding normal or shear stress values, at which transmission electron microscopy samples are taken, are given in Table 3.

The microstructure of the samples after fracture under uniaxial tensile loading was investigated on a scanning electron microscope, Zeiss Leo 1455 VP SEM. In order to obtain TEM foils, flat disks with a diameter of 3 mm and a thickness of approximately 500 μm were cut from the center of the deformation zone by wire eroding. These disks were mechanically thinned and polished from both sides to a thickness of 100 μm. The TEM-foils were electropolished with an electrolyte consisting of 120 mL 40% perchloric acid, 440 mL butoxyethanol and 440 mL 100% acetic acid in a Struers Tenupol Electropolisher 5. The microstructural investigations were performed on a JEOL JEM2010 Transmission Electron Microscope with a 200 kV electron gun.

Table 3. Relative deformation states of the TEM-samples in (true) stress-strain space (normal strain ε to normal Cauchy stress σ, resp. shear strain γ to shear Cauchy stress τ). ps, plane strain.

Material	Loading Case	Strain ε or γ $(-)$	Stress σ or τ (MPa)
DC06	(ps) tension	0.1	392
		0.2	449
	shear	0.1	136
		0.2	166
		0.3	184
AA6016	(ps) tension	0.1	255
		0.2	287
	shear	0.1	130
		0.2	153
		0.3	166

3. EXPERIMENTAL ANALYSIS OF MICROSTRUCTURAL PROPERTIES

3.1. Fracture Surfaces

The fracture surfaces indicate the nature of the plastic deformation in metallic materials. After the plastic deformation of the aluminum and steel samples up to the point of fracture under uniaxial tension, the features of the topology of the fracture surfaces were analyzed in detail. The corresponding fracture strains are $\varepsilon_{fr,\ DC06} = 0.554$ and $\varepsilon_{fr,\ AA6016} = 0.283$.

The analysis was carried out in three distinct parts (top, center and bottom) of the fracture surface. Figure 1 shows the resulting fractographs at the top and center for DC06 and AA6016. The fractographs of the bottom look similar to the one at the top and are skipped for the sake of brevity. The fractographs of the top and bottom show areas exhibiting facets of cleavage in the steel, as well as in the aluminum alloy (Figure 1a,c). Inclusions are clearly seen in the dimples on the fracture surface. In the case of alloys with the same lattice structure, the difference in the fracture toughness is manifested in the size of the dimples on the fracture surface, even if the lattice structure is changed [39]. Basically, all details of the fracture surface with a strongly curved pattern are cleavage features. Very fine bands are recognizable among the cleavage features exhibiting similarities to patterns observed for material fatigue under cyclic loading (fatigue striations).

The fracture surface's center shows a ductile fracture dominated by dimples due to microvoid coalescence. These dimples are of an equiaxed shape with a size of approximately 5 μm – 20 μm. The main factor accelerating the generation of dimples due to microvoid coalescence is the large preference of the material to from dimples rather than a different type of fracture. The fracture mechanism is further affected by external factors, such as stress conditions, as well as the strain rate. The pores emerge and grow at inclusions, which are clearly seen in the center of the fracture surface (Figure 1c,d). Additionally, the fracture surface of the steel sample DC06 exhibits deeper dimples in the vicinity of small dimples. Taylor [40] argues that these dimples induce local ductile fracture. In our aluminum samples, smaller parts of the second phase with a size of about 100 nm are observed within the dimples. The main precipitate in 6000 series Al-Mg-Si alloy systems are of type Mg_2Si; see e.g., [41,42]. Thus, for both materials, a mixed type of fracture is observed. Consequently, some regions in the fracture zone exhibiting ductile fracture originate in the coalescence of micro-voids, as well as separated grain facets, resulting in brittle fracture [40].

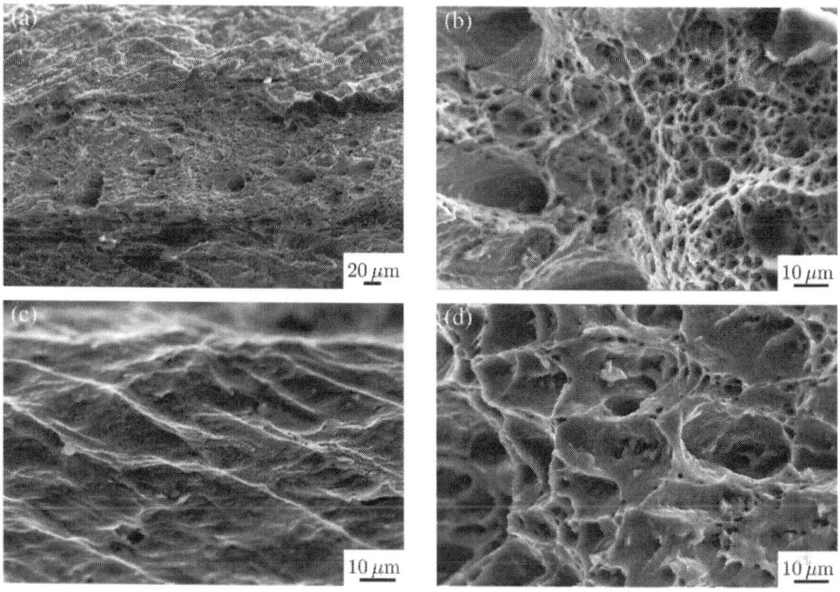

Figure 1. SEM fractograph of the top (**a**,**c**) and center (**b**,**d**) of the fracture surfaces of DC06 (**a**,**b**) and AA6016 (**c**,**d**) in uniaxial tension.

Although steel DC06 and aluminum AA6016 show certain differences as they feature different grain sizes and mechanical properties, the structures at the fracture surfaces, however, are similar. From our observation, we conclude that fracture behavior is similar in both materials, indicating that the underlying processes are comparable, as well. Next, the dislocation structures of both materials are investigated under different monotonic loading conditions to clarify the similarities and differences in the dislocation structure.

3.2. Microstructural Investigation

In the description of the dislocation microstructure, the following terms are used for microstructural elements: (i) cells are defined as regions of low dislocation density surrounded by lines of higher dislocation density; within the cells, the dislocations are homogeneously distributed; usually neighboring cells have misorientations from 0.2° to 0.8°; (ii) cell walls are regions of higher dislocation density surrounding the cells; the misorientation from the cell interior to the cell wall can be as large as 1°–1.5°; and (iii) cell-block boundaries are dense dislocation walls, which surround an area of dislocation cells; they are parallelepiped in shape and confine planar persistent dislocation structures.

3.2.1. IF-Steel DC06: TEM Image and Analysis

Figure 2a shows a TEM micrograph of the as-received material DC06. The initial texture is a γ-fiber texture with $\langle 111 \rangle$ direction oriented parallel to the sheet normal direction, which is characteristic for cold-rolled IF-steels [4]. As stated by Nesterova *et al.* [7], the orientation of the grains along the γ-fiber influences the dislocation distribution at shear strains up to 30%. The average grain size is determined as 20 μm, and the grain sizes range from 5 μm–60 μm, where the maximal and minimal grain size is only found rarely. The average deviation from the average grain size is very small. All specimens used in this work were manufactured from the same batch of material with average r-values (*i.e.*, anisotropy parameter): $r_{0°}$= 2.31,$r_{45°}$ = 1.95, $r_{90°}$ = 2.77.

In uniaxial tension, the material tends to form cell-block boundaries initially predominantly associated with {110} glide systems. As deformation continues, cell-block boundaries belonging to {112} systems are observed. After 15% strain, cell-block boundaries oriented on {123} planes appear, as well. In general, the cell-block boundaries are observed after 5% strain, but they are more pronounced at larger strains (e.g., 20%). At strains between 5% and 8%, cells and planar persistent structures coexist. Additionally, between 5% and 12% tensile strain, dislocations tangles are observed, as well. Our analysis indicates that these appear in certain parts of the sample in this strain range, where other tangles seem to disappear simultaneously. After 12% tensile strain, nearly all dislocation tangles are incorporated into cell walls. With increasing deformation, the misorientation angle between cell interior and cell-block boundaries increases. Finally, this results in sharp boundaries referred to as blade-like plates on a larger scale (*cf.* Rybin [43]). The rolling texture increased with tensile loading, as reported in [4], where a strengthening of the γ_2-component (see [7] for terminology) is found to accumulate around $\phi_2 = \{0°, 60°\}$.

Consistent with observations of Bouvier *et al.* [34], during plane strain tension, the degree of irregularity of the dislocation microstructure increases compared to uniaxial tensile loading (see Figure 2b,c). Cell block boundaries start to form analogously. Under plane strain tension, a second group of cell-block boundaries oriented in another direction of the family ⟨111⟩ is observed more often. The cell-block boundaries exhibit more curvature and are more diffuse than in uniaxial tension. The electron-microscopical investigations show that up to a tensile strain of 10%, grain sizes of 10 μm − 12 μm are observed. In deformed grains, cell walls are formed by kinked arrangements of dislocations. In these regions, dislocations rearranging into sub-grain-like boundaries are observed. They are referred to as dense dislocation walls. In elongated grains with a high aspect ratio, the grains are fragmented into subgrains of about 1 μm in width. The boundaries of these subgrains appear as dislocation walls and networks, which sometimes form subgrain boundaries with a fringe contrast. The dense dislocation walls and networks are inclined 15°–20° to the macroscopic tensile axis. As reported by Clausmeyer *et al.* [4], at the same normal strain, the dislocation structure under uniaxial tension consists of twice as much cells as under plane strain tension. With increasing deformation up to 20% strain, the cell size decreases and the dislocation density in areas of tangles and accumulations increases. This is accompanied by a tendency to arrange parallel subboundaries. The grain size also decreases to 6 μm–8 μm. Dense dislocation walls are parallel to traces of {110} planes, which is the rolling plane (trace of plane). The rolling texture increases under plane strain tension with a strengthening of the γ_2-component as under uniaxial tension.

After simple shear deformation of 10% strain, grain sizes of up to 10 μm are observed analogous to tensile loading. Figure 2d shows a typical pattern for the dislocation structure after 10% shear strain. The development of two families of dense dislocation walls within a single grain are observed (trace of plane). These dislocation walls incline 5°–8°in the longitudinal plane toward the shear direction. An important requirement for the creation of a certain texture or microstructure is the interaction of several slip systems [44]. With shear deformation up to 20% strain, the grain size decreases slightly (e.g., to 9.4 μm), while the dislocation density in the grains, as well as on the grain boundaries increases. Dense dislocation walls are inclined 10°–25° with respect to the shear direction. Between 20% and 30% shear strain, grain sizes up to 10 μm are measured and the intensity of dislocation networks increases. Additionally, grains with single dislocations are found. The forming cell blocks show a rectangular shape (*i.e.*, aspect ratio close to one) compared to the quadrilateral shape after tensile loading. Under shear around 50%, more cells have developed compared to tensile loading at the equivalent amount of strain. The formation of micro shear bands starts at shear strains of about 30%. They enclose angles of 35°–45° with the plane of deformation $\{110\}$. Noteworthy is the fact that their formation has no relation with pre-existing dislocation cells. Figure 2e,f show the dislocation structure of a micro-band in the specimen. At 20% strain, the nucleation of a micro-band can be anticipated, where the full one is created at larger strain values. The boundaries of the band are dense dislocation walls and are approximately parallel to $\{110\}$ planes. Each band consists of two families of dislocation walls (oriented 35° to $\langle 110 \rangle$. One family parallel to the longitudinal direction of the grain (see Figure 2e) has a distance of about 50 nm between single dislocation walls compared to the family that is significantly smaller than the distance of 100 nm–200 nm between single dislocation walls. The initiation of micro-bands often starts at large precipitates or grain boundaries. Precipitates and grain boundaries serve as dislocation sources contributing to a higher dislocation density on a single slip system. This might result in the formation of dense dislocation walls and, finally, to the appearance of micro-bands. Micro-bands, not initiated by large precipitates or grain boundaries, can be found for small carbide particles of about 5 nm–50 nm. Figure 2f shows two families of micro-bands within a grain at 30% shear deformation. The micro-bands are 100 nm–400 nm thick. The area of micro-bands increases and the morphology becomes similar to an arrangement of blade-like plates of dense dislocation walls. In Figure 2f, the micro-band walls are located parallel to $\{101\}$ planes. Localization of deformation in narrow bands (deformation bands) starts when parts of a single crystal start to reorient due to several reasons, such as anisotropy, texture and strain rate. This process is called orientation split-up. Two families of cell-block boundaries are predominant, either parallel or orthogonal to the shearing direction. At large amounts of shear, such as 50%, in some grains, both families are found at the same time. For large amounts of deformation, misorientations from 0.8° up to 5° are observed within the cell-block boundaries, with most misorientations being smaller than 3°.

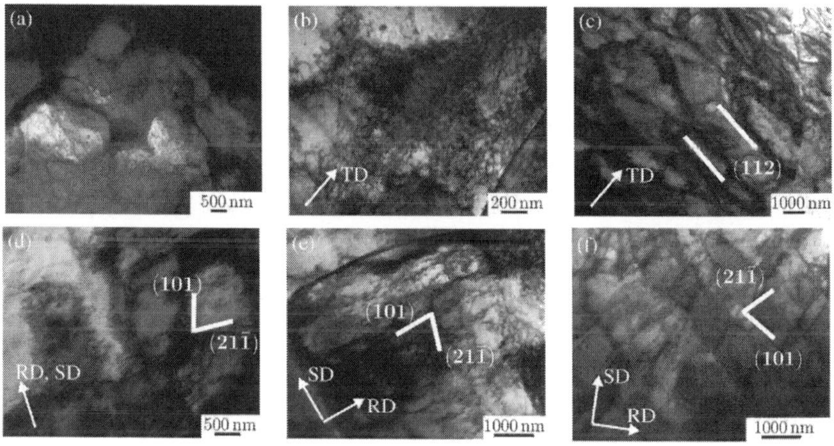

Figure 2. TEM images of dislocation structures of DC06: (**a**) as-received material; after plane strain tension up to 10% (**b**) and up to 20% (**c**) for tension in the rolling direction (RD); after 10% (**d**), 20% (**e**) and 30% (**f**) simple shear loading where shear direction (SD) is orthogonal to the RD. The red arrows in (**e**) indicate a family of dislocation walls parallel to the longitudinal direction of the grain center with a distance of about 50 nm (which is significantly smaller than the distance of 100–200 nm between single dislocation walls of other families).

3.2.2. Aluminum Alloy AA6016-T4: TEM-Image and Analysis

An analogous investigation was performed on the fcc aluminum alloy AA6016-T4, and the results are correlated to the previous one from

DC06. Figure 3a shows a TEM micrograph of the as-received material AA6016-T4, which exhibits a grain size of approximately 20 μm–25 μm and average r-value $r_{0°} = 0.63$, $r_{45°} = 0.41$, $r_{90°} = 0.77$ [38]. The electron-microscopical investigation of the alloy after plane strain tension after 10% deformation shows that dislocation walls and dislocation nets are observed in the microstructure. The dislocation walls are inclined about ±10° to the loading direction. In addition, the number of grains exhibiting a cell structure is significantly higher than in DC06. With increasing deformation up to 20%, the dislocation density is increased (see Figure 3b). Most dislocation walls are now inclined about ±40° to the loading axis. The formation of dislocation structures is intensified. Often, two families of dislocation walls are found in the microstructure, namely on $\{111\}$ in two directions of the family $\langle 110 \rangle$. These walls are parallel to traces of $\{111\}$ slip planes and not to the tensile direction. Changes in the degree of deformation of the aluminum alloy sample result in high dislocation densities, dislocation clusters redistribution inside the cell-structure and cell boundaries similar to those observed in DC06.

After a plastic deformation of 10% shear, the grains feature an approximate size of 1 μm and the dislocation structure is identified as a dislocation net (see Figure 3d). In addition to the formation of dislocation walls and dislocation nets, single dislocations are observed. As in the case of plane strain tension, dislocation nets in the longitudinal planes are inclined about ±10° to the rolling direction. With increasing shear deformation up to 20%, grains are of size of about 6 μm–8 μm, which probably increased due to recrystallization effects. Furthermore, the dislocation density increases. The formation of dislocation walls and dislocation nets are intensified. Two families of nets are seen in Figure 3e; however, in most of the cases investigated, only one family is observed. This family of net is formed with an inclination angle of 35° with respect to the macroscopic shear direction. However, in some grains, micro-bands begin to form (see Figure 3f) where two families of micro-bands are observed within a grain, as well. The thickness of the micro-bands is identified as about 50 nm. The micro-band walls are parallel to $\{110\}$ planes. The trace of the second family of micro-bands are parallel to the $\{101\}$ planes.

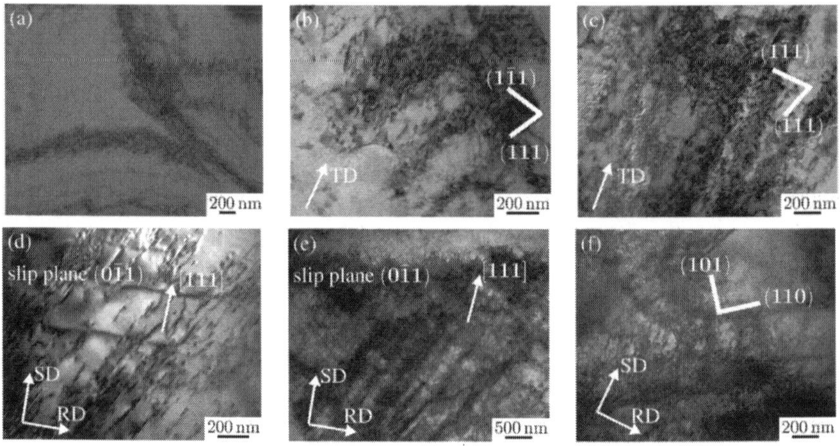

Figure 3. TEM images of dislocation structures of AA6016: (**a**) as-received material; after plane strain tension up to 10% (**b**) and up to 20% (**c**) for tension in the RD; after 10% (**d**), 20% (**e**) and 30% (**f**) simple shear loading, where the SD is orthogonal to the RD.

4. DISCUSSION

The characteristic features of the dislocation structure and its dynamic evolution under the two loading conditions plane strain tension and simple shear are summarized in Table 4 for aluminum alloy AA6016-T4 and IF-steel DC06. In the uniaxial tension tests, the dislocation structures develop more intensively and quickly compared to the plane strain test. In DC06, dislocation cells start to develop in multiple grains with a favorable Schmid factor at 5% strain in uniaxial tension test, whereas in plane strain tension, this is only seen at approximately 10% strain. In the case of AA6016, the same, but less pronounced behavior is observed.

Plane strain tension in DC06 leads to a less intensive deformation at the micro level compared to uniaxial tensile loading. Nevertheless, at strains higher than 10%, a detailed measurement of dislocation densities is not possible due to the fact that dislocation walls and cells with high dislocation densities have already developed. The cells show fewer dislocations in the interior and are formed in a polygonal shape, with most being rectangular (cell blocks). These cells are larger than the created fragments in uniaxial tension at the same strain level. The intensity of plastic deformation is larger under tension compared to shear, because under tension, more pronounced and intense dislocation structures appear. Already at low strains (*approx.* 10%), the formation of dislocation walls is observed. The number of cells and the aspect ratio *a/b* of the cells are compared in Table 5 for DC06 and AA6016-T4 under plane strain tension and shear at various deformation states. Here,*a* denotes the size in or close to the direction of loading and *b* denotes the size perpendicular to that. Under simple shear, the characteristic dislocation cell shape is close to a quadratic shape.

Table 4. Characteristics of dislocation structures under plane strain deformation in aluminum alloy AA6016-T4 and steel DC06.

Loading Conditions	Strain Level	Characteristics of Dislocation Structure	
		Aluminum Alloy AA6016-T4	Steel DC06
plane strain tension	0	single dislocations, accumulations of dislocations	single dislocations, cell structure
	0.1	single dislocations, start of formation of dislocation walls	cell structure and start of formation of dislocation walls
	0.2	formation of dislocation cells parallel to two directions	cell structure, interaction between dislocation walls, micro-shear bands
shear	0.1	single dislocations, start of formation of dislocation walls parallel to two directions, cell structures	single dislocations, dislocation pile-up
	0.2	formation of dislocation cells, dislocation walls	dislocation pile-up, cell structures
	0.3	on basis of the dislocation walls form micro-shear bands	dislocation pile up, cell structure and start of formation of dislocation walls

In AA6016-T4, the formation of dislocation structures is occurring at a higher intensity in the plane strain tension test compared to the one in shear. Already at low strains, dislocation walls of high dislocations (approximately 10^{14} m/m^3 to 3×10^{14} m/m^3; see [45] and [46] for details about dislocation density calculations from TEM images) are observed in plane strain tension. Under shear loading, the dislocations form quadratically-shaped dislocation walls at strains up to 20%, consisting partially of single dislocations. These cell walls are smaller (250 nm–500 nm for plane strain tension, 150 nm–300 nm for shear) and have a lower dislocation density (approximately 2×10^{13} m/m^3 to 5×10^{13} m/m^3) compared to the ones under plane strain tension. A further strain increase leads to the formation of micro-shear bands, which are formed on the basis of the previously-formed dislocation cell walls. The cell shapes under plane strain tension are rectangular with an aspect ratio of 1:2. Under simple shear, the cells formed first have an aspect ratio of 1:3, and with increasing deformation, the cells become quadratic.

Table 5. Microstructural properties of the dislocation structures under plane strain tension, resp. shear loading, in aluminum AA6016-T4 and steel DC06 at various strain levels.

Loading Conditions	Strain	Number of Cells Per Grain		Aspect Ratio a/b of Cells		Orientation of Cells to Loading Direction	
		AA6016-T4	DC06	AA6016-T4	DC06	AA6016-T4	DC06
plane strain tension	0.1	10 to 20	2 to 6	1:2 to 1:1	1:2 to 1:3	10°	10°–15°
	0.2	25 to 30	6 to 8	1:2	1:1 to 1:4	35°–40°	15°–20°
shear	0.1	20	-	1:3 to 1:2	1:1, rarely 1:2	3°–10°	5°–8°
	0.2	30 to 40	5	1:2	1:1	35° 40°	10° 25°
	0.3	50 to 60	15	1:1	1:1, rarely 1:2	35°–40°	30°–35°

Comparing the results of the two materials with different lattice structures (bcc and fcc), it is noted that the deformation in DC06 is more intense than in AA6016-T4. A classification of the microstructural features of both materials is suggested inFigure 4. The focus is on a qualitative description of the dislocation structure without an exact quantification of the dislocation density. In general, differences in the intensity of the formation of dislocation structures are observed in plane strain and shear loading. The investigation of the dislocation densities in this work is based on two to three measurements at each strain level. This corresponds to the analysis of ten to twelve pictures of the dislocation structure tilting differently from the specimen holder.

By comparing the intensity in plane strain tension of the formation of dislocation structures between both materials, the following observations can be made:

- In the as-received material, as well as at low strain levels (0%–5%), single dislocations and accumulations of dislocations are observed in aluminum, whereas in steel, already dislocation cells are starting to form.

- An increase of the strain level up to 10% leads to the formation of dislocation walls in both materials.
- A further increase in strain leads to differences in the dislocation structure evolution: in aluminum, an orientation/direction-dependent formation of dislocation cells is occurring; in steel, a strong interaction between the dislocation walls and the formation of micro-shear bands are observed.

The formation of the dislocation structures under shear in aluminum and steel show very different tendencies compared to the ones under plane strain:

- In aluminum, the formation of dislocation cells starts earlier than in steel, even for strain levels smaller than 0.1.
- An evolution of the dislocation structure is not based on the intensive formation of dislocation cells and dislocation walls. Dislocation walls in combination with micro-shear bands in aluminum and in steel form due to the formation of strongly-branched dislocation structures.

The comparison of the hardening in both materials shows that the hardening behavior is similar, although differences are seen in the intensity of the microstructural processes. The dislocation structures in DC06 and AA6016-T4 show certain differences; however, the dynamics in the hardening behavior is similar. This indicates that the principles in the plastic deformation are rather defined by the form and distribution of dislocation structures than by the glide mechanism alone. Both materials exhibit a monotonic hardening behavior, where hardening saturates with increasing amounts of strain.

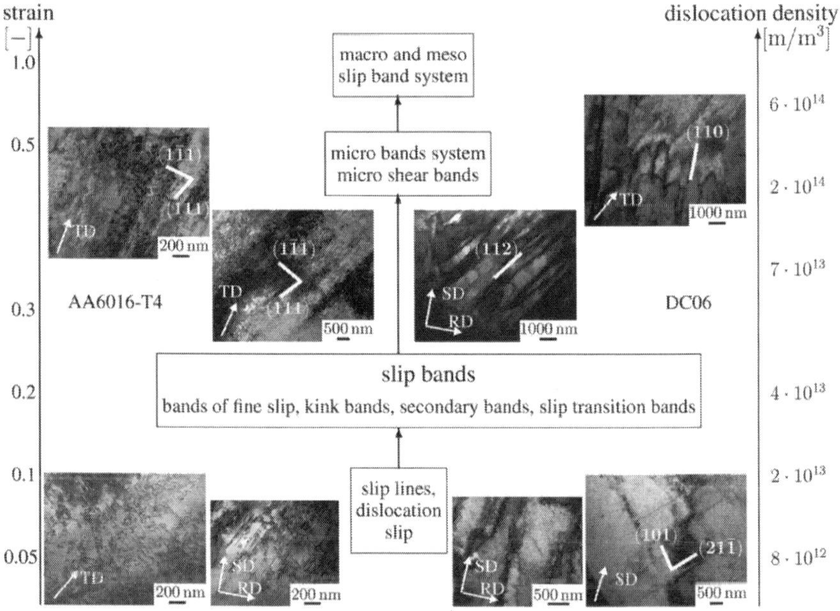

Figure 4. Classification of microstructural elements in aluminum alloy AA6016-T4 and steel DC06 observed at different levels of deformation. These are correlated with the observed dislocation densities.

5. SUMMARY AND OUTLOOK

On the basis of plane strain tension and simple shear tests, the evolution of the grain and dislocation microstructures have been investigated for two materials with different lattice structures, namely an interstitial free steel (DC06) and a 6000 series aluminum alloy (AA6016-T4). Within this work, the microstructural changes and features at different strain states were characterized and compared for both materials to provide information on the active mechanism at the microscale governing the macroscopic response. The observations suggest that the plastic deformation mechanisms are rather defined by the form and distribution of dislocation structures rather than by the glide mechanism alone. Overall, the characterization of dislocation structure revealed that the deformation in DC06 is more intense than in

AA6016-T4. Different tendencies in the dislocation formation were obtained under plane strain tension compared with shear loading for both materials, which has been discussed in detail. Further, it was found that the fracture behavior of DC06 and AA6016 is similar under uniaxial tensile loading, indicating that the processes are similar, as well.

As a next step, these results can be used to motivate and validate new micromechanical-based numerical models. A thoughtful material model for the macroscopic response requires the modeling of the microstructural mechanisms resulting in dislocation patterns. Therefore, a detailed investigation for different loading histories, as presented in this work, provides a good basis. Only such detailed local analyses allow for the determination of differences in the behavior of materials and provide conclusive results regarding the responsible physical mechanisms.

ACKNOWLEDGMENTS

Financial support provided by the German Science Foundation (DFG) under contract PAK 250 (TP4, TP5) is greatly acknowledged. This publication was supported by the German Research Foundation (DFG) and Hamburg University of Technology in the funding program 'Open Access Publishing'. The materials investigated were provided by ThyssenKrupp Steel Europe AG (DC06) and Novelis (Sierre, Switzerland) (AA6016-T4). The tests at Twente University were performed by Till Clausmeyer (Institute of Mechanics, TU Dortmund, Germany), which is gratefully acknowledged.

AUTHOR CONTRIBUTIONS

Gregory Gerstein performed the uniaxial tensile tests and investigated all samples on the scanning and transmission electron microscope. All authors discussed the results. Gregory Gerstein, Benjamin Klusemann and Swantje Bargmann wrote the article.

CONFLICTS OF INTEREST

The authors declare no conflict of interest.

REFERENCES

1 Von Mises, R. Mechanik der festen Körper im plastisch-deformablen Zustand. In *Nachrichten von der Gesellschaft der Wissenschaften zu Göttingen, Mathematisch-Physikalische Klasse*; Göttinger Digitalisierungszentrum: Goettingen, Germany, 1913; pp. 582–592. (In German)

2 Hill, R. A theory of the yielding and plastic flow of anisotropic metals. *Proc. Roy. Soc. Lond.* **1948**, *193*, 281–297.

3 Barlat, F.; Gracio, J.J.; Lee, M.G.; Rauch, E.F.; Vincze, G. An alternative to kinematic hardening in classical plasticity.*Int. J. Plast.* **2011**, *27*, 1309–1327.

4 Clausmeyer, T.; Gerstein, G.; Bargmann, S.; Svendsen, B.; Boogaard, A.; Zillmann, B. Experimental characterization of microstructure development during loading path changes in bcc sheet steels. *J. Mater. Sci.* **2013**, *48*, 674–689.

5 Fernandes, J.V.; Schmitt, J.H. Dislocation microstructures in steel during deep drawing. *Philos. Mag. A* **1983**, *48*, 841–870.

6 Ghosh, A.K.; Backofen, W.A. Strain hardening and instability in biaxially stretched sheets. *Metall. Trans.* **1973**, *4*, 1113–1123.

7 Nesterova, E.; Bacroix, B.; Teodosiu, C. Experimental observation of microstructure evolution under strain-path changes in low-carbon IF steel. *Mater. Sci. Eng. A* **2001**, *309*, 495–499.

8 Peeters, B.; Kalidindi, S.R.; Teodosiu, C.; Houtte, P.V.; Aernoudt, E. A theoretical investigation of the influence of dislocation sheets on evolution of yield surfaces in single-phase B.C.C. polycrystals. *J. Mech. Phys. Solids* **2002**, *50*, 783–807.

9 Rauch, E.F.; Schmitt, J.H. Dislocation substructures in mild steel deformed in simple shear. *Mater. Sci. Eng.* **1989**, *441*, 441–448.

10 Thuillier, S.; Rauch, E. Development of micro-bands in mild steel during cross loading. *Acta Metall. Mater.* **1994**, *42*, 1973–1983.

11 Wilson, D.; Bate, P. Influences of cell walls and grain boundaries on transient responses of an IF steel to changes in strain path. *Acta Metall. Mater.* **1994**, *42*, 1099–1111.

12 Yin, Q.; Zillmann, B.; Suttner, S.; Gerstein, G.; Biasutti, M.; Tekkaya, A.E.; Wagner, M.F.X.; Merklein, M.; Schaper, M.; Halle, T.; *et al.* An experimental and numerical investigation of different shear test configurations for sheet metal characterization. *Int. J. Solids Struct.* **2014**, *51*, 1066–1074

13 Van Riel, M. Strain Path Dependency in Sheet Metal-Experiments and Models. Ph.D. Thesis, Twente University, Enschede, The Netherlands, 2009.

14 Bauschinger, J. Über die Veränderung der Elasticitätsgrenze und des Elasticitätsmoduls verschiedener Metalle. *Zivilingenieur* **1881**, *27*, 289–348.

15 Zillmann, B.; Härtel, M.; Halle, T.; Lampke, T.; Wagner, M. Flow behavior of automotive aluminum sheets during in-plane uniaxial and biaxial compression loading. *Steel Res. Int.* **2011**, *1*, 691–694.

16 Merklein, M.; Biasutti, M. Forward and reverse simple shear test experiments for material modeling in forming simulations. In Proceedings of the 10th International Conference on Technology of Plasticity, ICTP 2011, Aachen, Germany, 25–30 September 2011; Wiley: Hoboken, NJ, USA, 2011; Volume 1, pp. 702–707.

17 Hasegawa, T.; Yakou, T.; Karashima, S. Deformation behaviour and dislocation structures upon stress reversal in polycrystalline aluminium. *Mater. Sci. Eng.* **1975**, *20*, 267–276.

18 Anand, L.; Gurtin, M.E.; Lele, S.P.; Gething, C. A one-dimensional theory of strain-gradient plasticity: Formulation, analysis, numerical results. *J. Mech. Phys. Solids* **2005**, *53*, 1789–1826.

19 Bardella, L. A deformation theory of strain gradient crystal plasticity that accounts for geometrically necessary dislocations. *J. Mech. Phys. Solids* **2006**, *54*, 128–160.

20 Bargmann, S.; Svendsen, B.; Ekh, M. An extended crystal plasticity model for latent hardening in polycrystals.*Comput. Mech.* **2011**, *48*, 631–645.

21 Evers, L.P.; Brekelmans, W.A.M.; Geers, M.G.D. Scale dependent crystal plasticity framework with dislocation density and grain boundary effects. *Int. J. Solids Struct.* **2004**, *41*, 5209–5230.

22 Holmedal, B.; van Houtte, P.; An, Y. A crystal plasticity model for strain-path changes in metals. *Int. J. Plast.* **2008**, *24*, 702–707.

23 Husser, E.; Lilleodden, E.; Bargmann, S. Computational modeling of intrinsically induced strain gradients during compression of c-axis oriented magnesium single crystal. *Acta Mater.* **2014**, *71*, 206–219.

24 Kalidini, S.R.; Bronkhorst, C.A.; Anand, L. Crystallographic texture evolution during bulk deformation processing of fcc metals. *J. Mech. Phys. Solids* **1992**, *40*, 537–569.

25 Marin, E.B.; Dawson, P.R. On modelling the elasto-viscoplastic response of metals using polycrystal plasticity.*Comput. Methods Appl. Mech. Eng.* **1998**, *165*, 23–41.

26 Clausmeyer, T.; van den Boogaard, T.; Noman, M.; Gershteyn, G.; Schaper, M.; Svendsen, B.; Bargmann, S. Phenomenological modeling of anisotropy induced by evolution of the dislocation structure on the macroscopic and microscopic scale. *Int. J. Mater. Form.* **2011**, *4*, 141–154.

27 Ohno, N.; Okumura, D. Higher-order stress and grain size effects due to self-energy of geometrically necessary dislocations. *J. Mech. Phys. Solids* **2007**, *55*, 1879–1898.

28 Roll, K.; Wiegand, K.; Hora, P.; Manopulo, N.; Clausmeyer, T. Benchmark 2—Influence of drawbeads on the springback behaviour (S-RAIL-08). In Proceedings of the 7th International Conference and Workshop on Numerical Simulation of 3D Sheet

Metal Forming Processes, Numisheet, Interlaken, Switzerland, 1–5 September 2008; Numisheet: Melbourne, Australia, 2008; pp. 53–111.

29 Barthel, C.; Klusemann, B.; Denzer, R.; Svendsen, B. Modeling of a thermomechanical process chain for sheet steels.*Int. J. Mech. Sci.* **2013**, *74*, 46–54.

30 Behrouzi, A.; Soyarslan, C.; Klusemann, B.; Bargmann, S. Inherent and induced anisotropic finite visco-plasticity with applications to the forming of DC06 sheets. *Int. J. Mech. Sci.* **2014**, *89*, 101–111.

31 Rabahallah, M.; Balan, T.; Bouvier, S.; Bacroix, B.; Barlat, F.; Chung, K.; Teodosiu, C. Parameter identification of advanced plastic strain rate potentials and impact on plastic anisotropy prediction. *Int. J. Plast.* **2009**, *25*, 491–512.

32 Hu, Z. Work-hardening behavior of mild steel under cyclic deformation at finite strains. *Acta Metall. Mater.* **1994**, *42*, 3481–3491.

33 Teodosiu, C.; Hu, Z.H. Evolution of the intragranular microstructure at moderate and large strains: Modelling and computational significance. *Proc. Numiform* **1995**, *95*, 173–182.

34 Bouvier, S.; Gardey, B.; Haddadi, H.; Teodosiu, C. Characterization of the strain-induced plastic anisotropy of rolled sheets by using sequences of simple shear and uniaxial tensile tests. *J. Mater. Process. Technol.* **2006**, *174*, 115–126.

35 *Cold Rolled Low Carbon Steel Flat Products for Cold Forming*; Technical Report for Arcelormittal: Luxembourg City, Luxembourg, 2006.

36 *Aluminium and Aluminium Alloys—Chemical Composition and form of Wrought products—Part 3: Chemical Composition and form of Products*; European Standards: Pilsen, Czech, 2007; EN 573-3:2007.

37 Van Riel, M.; van den Boogaard, A. Stress-strain responses for continuous orthogonal strain path changes with increasing sharpness. *Scr. Mater.* **2007**, *57*, 381–384.

38 Zillmann, B.; Clausmeyer, T.; Bargmann, S.; Lampke, T.; Wagner, M.X.; Halle, T. Validation of simple shear tests for parameter identification considering the evolution of plastic anisotropy. *Tech. Mech.* **2012**, *32*, 622–630.

39 Fellows, J.A.; Boyer, H.E. *Metals Handbook. Fractography and Atlas of Fractographs*; American Society for Metals: Materials Park, OH, USA, 1974; Volume 9.

40 Taylor, G. The mechanism of plastic deformation of crystals. Part II. Comparison with observations. *Proc. R. Soc. Lond. A* **1934**, *145*, 388–404.

41 Elangovan, K.; Balasubramanian, V. Influences of post-weld heat treatment on tensile properties of friction stir-welded AA6061 aluminum alloy joints. *Mater. Charact.* **2008**, *59*, 1168–1177.

42 Wanderka, N.; Schiffmann, R.; Banhart, J. Characterization of precipitates in aluminium-based alloy AW 6016. *Surf. Interface Anal.* **2007**, *39*, 221–226.

43 Rybin, V. *Severe Plastic Deformations and Fracture of Metals*; Metallurgiya: Moscow, Russia, 1986. (In Russian)

44 Kalidini, S.R.; Bronkhorst, C.A.; Anand, L. Crystallographic texture evolution during bulk deformation processing of fcc metals. *J. Mech. Phys. Solids* **1992**, *40*, 537–569.

45 Hirsch, P.B.; Howie, A.; Nicholson, R.B.; Pashley, D.W.; Whelan, M.J. *Electron Microscopy of Thin Crystals*; Butterworths: London, UK, 1965.

46 Bailey, J.E.; Hirsch, P.B. The dislocation distribution, flow stress, and stored energy in cold-worked polycrystalline silver. *Philos Mag.* **1960**, *5*, 485–497.

CHAPTER 3

Electrochemical Behavior and Corrosion Study of Electrodeposits of Zn and Zn-Fe-Co on Steel

Conceição A. M. Dutra*, Eduardo N. Codaro, Roberto Z. Nakazato

Departamento de Física e Química, Universidade Estadual Paulista, Campus de Guaratinguetá, Guaratinguetá, Brazil.

ABSTRACT

From industrialized baths, free of cyanide, the corrosion behavior of electrodeposits of zinc and zinc alloys was studied by means of electrochemical tests in aerated solution of 3.5% (0.6 M) NaCl at pH 8.2. In literature, several studies are found about zinc coatings and zinc alloys, for example, Zn-Ni and Zn-Co, nevertheless there is little about the ternary alloy Zn-Fe-Co. The Fe presence in the alloy results in a good adhesion to the substrate and allows application of these materials at higher temperatures. The electrochemical tests were carried out by obtaining open circuit potential curves with immersion time, potentiodynamic polarization curves and cyclic voltammetry. From the obtained results, the large potential differences observed between the steel and the electrodeposits showed that the last protect the substrate, acting as a sacrifice metal. The tests disclosed similar behaviors in both the current densities and the corrosion potential for electrodeposits of Zn and Zn-Fe-Co. After chromate passivation process, a significant decrease in corrosion density was noted for Zn and when the system was de-aerated there was change in the cathodic process mechanism.

KEYWORDS

Corrosion; Electrodeposition; Zn; Zn Alloy

1. INTRODUCTION

Zinc gives galvanic protection to steel by presenting a good accomplishment in most atmospheric conditions and also good mechanical and corrosive characteristics. In relation to electrogalvanised layers, zinc alloys have been commonly more used due to their low cost and their resistance to the corrosion process, which can triple, making them significantly more efficient [1]. Zinc electrodeposition with metals of Fe group, besides cost advantage, it is classified as anomalous [2]; these coatings present protection mechanisms by barrier associated with cathodic protection [3,4].

The increasing use in the industries has caused a great progress in technology of coatings based on zinc. In recent years, particularly in automotive industry, it has increased the requirements and the demand of coatings, which are alternative to zinc and cadmium for steel protection. It has resulted in the development of zinc alloys with metals of group VIII B, such alloys present superior properties of corrosion resistance and low toxicity. Furthermore, characteristics of these coatings can be improved by post-treatment.

Chemical conversion processes with chromates are widely applied in metals passivation. Chromatization is used as a more effective inhibiting process, although the hexavalent chromium presents very toxic effects. As a result, oxy-anions similar to chromate, such as molybdate, tungstate and vanadate, are studied as alternatives. The objective of this paper is to study the electrochemical behavior of Zn and Zn-Fe-Co electrodeposits upon steel in 3.5% NaCl, by analyzing the addition effect of CrO_4^{2-}, MoO_4^{2-} and WO_4^{2-} 0.05 M.

2. MATERIALS AND METHODS

Using as substrate SAE 1010 steel plates 0.75 mm thick, the working electrode was prepared with pure Zn electrodeposits (NCZ ZINCROLYTE 191ª) and Zn-Fe-Co (NCZ ZINCROLYTE 191 Fe/Coª) in commercial baths based on chlorides, cyanide-free and industrially obtained with a thickness of around 10 μm. During immersion of the samples, a current density from 2 to 4 $A \cdot dm^{-2}$ was applied with a temperature of about 25°C. All electrodeposition procedures were obtained in a laboratory of Electronics Enthone Brazil Ltda, which are properties and trademarks of Cookson Electronics Brazil Ltda. After electrodeposition procedure, a part of the sample was passivated by chemical conversion of blue chromate (PERMAPASS 3080ª). With a potentiostat EG&G PAR 283 support, controlled by programs Softcorr III and Power Suite, interfaced by GPIB plate to a computer for control and data processing, electrochemical tests were conducted in a flat corrosion cell with a capacity of 250 mL, borosilicate glass with compartments for platinum counter electrode, working electrode and reference electrode Ag/AgCl, KClsat (+197 mV vs. NHE). The used electrolyte was a 3.5% NaCl, pH 8.2, prepared with analytical grade reagent and distilled water, with and without group VIB oxy-anions presence (CrO_4^{2-}, MoO_4^{2-} and WO_4^{2-}) at a concentration of 0.05 M, at 25°C.

3. RESULTS AND DISCUSSION

3.1. Measurements of Open Circuit Potential (Eoc)

During 600 minutes without passing any electric current, pure electrodeposits of Zn and Zn-Fe-Co were analyzed by using 3.5% NaCl, pH 8.2, with and without chromate treatment and addition of oxy-anions. Compared to pure Zn, samples of Zn-Fe-Co, both for untreated and passivated, corrosion potentials remain close. When the pure Zn electrodeposits are treated by chromatization, it is noted that, in a general way, the potentials obtained in open-circuit are shifted to more positive values than those obtained for zinc without treatment, indicating that in chromatization procedure, it occurs stable species formation, which protect the material from a corrosive chloride attack. With oxy-anions addition to chloride, potential values variations occurred for Zn coating **Figure 1**(a). The corrosion potentials are shifted to more positive values in presence of all studied oxy-anions. This corrosion potential increase means a higher stability of the formed product. In **Figure 1**(b), CrO_4^{2-} and MoO_4^{2-} ions caused a corrosion potential increase, indicating a positive effect of inhibitors on zinc alloy protection.

WO_4^{2-} addition in the solution shifted the alloy potential to more negative values, showing that it has not any inhibitory effect from corrosion to the alloy, which means a more active surface.

Factors, such as pH and oxygen presence, must be taken into account in the Fe-Zn-electrolyte system, because it has a very complex behavior. During corrosion, zinc alloy surface on steel in chloride solution, even in low concentration, shows pH variation. pH is high in remote areas of the layer, where the mechanism proposed to explain this increase results from the zinc oxidation process Equation (1) and from the reduction parallel reaction of the dissolved oxygen in aqueous medium Equation (2); oxygen is reduced by forming OH^- ions, whereas Equation (3) is the predominant overall reaction. Next to the layer, pH becomes intermediate and in a region close to zinc layer it is low, being this decrease due to the metal hydrolysis reaction, where H^+ is simultaneously produced Equation (4).

$$Zn_{(s)} \rightarrow Zn^{2+}{}_{(aq)} + 2e^- \qquad (1)$$

$$\frac{1}{2}O_{2(g)} + H_2O_{(l)} + 2e^- \rightarrow 2OH^-{}_{(aq)} \qquad (2)$$

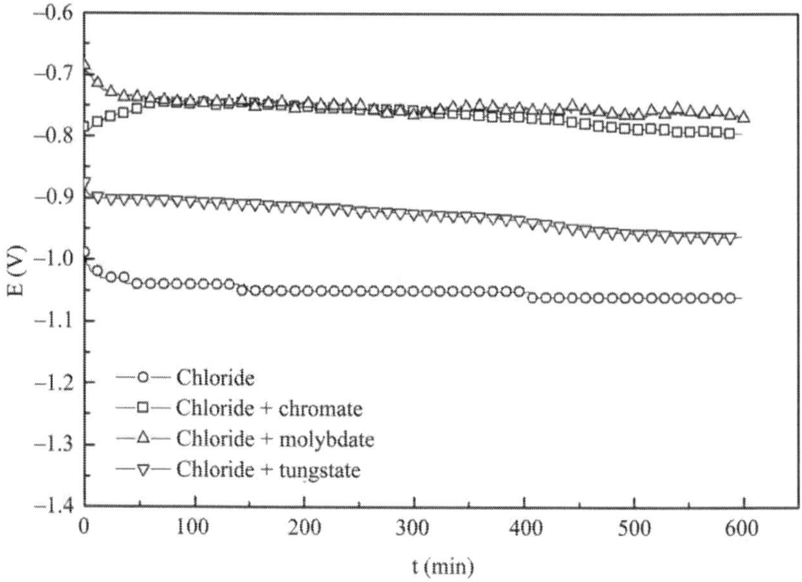

(a)

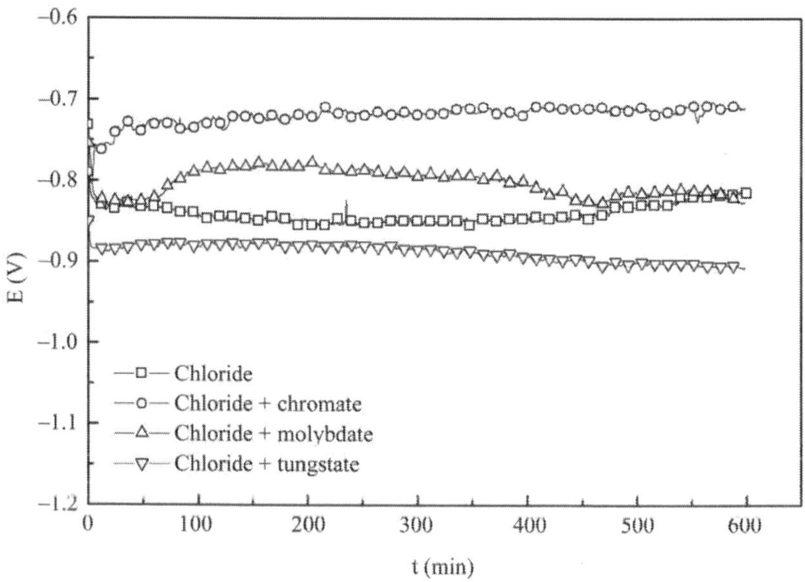

(b)

Figure 1. Open circuit potential curves in 3.5% NaCl solution, pH 8.2 and in presence of different oxy-anions 0.05 M, for electrodeposits: (a) Zn and (b) Zn-Fe-Co.

$$2Zn_{(s)} + O_{2(g)} + 2H_2O_{(l)} \rightarrow 2Zn(OH)_{2(s)} \quad (3)$$

$$Zn^{2+}{}_{(aq)} + H_2O_{(l)} \rightarrow ZnOH^{+}{}_{(aq)} + H^{+}{}_{(aq)} \quad (4)$$

Just above the zinc layer, in region containing low pH value, a high Zn^{2+} concentration is developed. Conversely, in the most distant regions of zinc layer, Zn^{2+} does not exist. During corrosion, due to concentration gradients of H^+, OH^- and Zn^{2+}, solutions are mixed, due to continuous H^+ production by releasing Zn^{2+}. These are the processes that possibly occur both on zinc and alloy surface, besides the anodic process of dissolution on zinc surface by chloride and passivation caused by oxygen reduction Equations (5) and (6) [5].

$$5Zn^{2+}{}_{(aq)} + H_2O_{(l)} + 8OH^{-}{}_{(aq)} + 2Cl^{-}{}_{(aq)} \rightarrow Zn_5Cl_2(OH)_8 \cdot H_2O_{(s)} \quad (5)$$

$$Zn^{2+}{}_{(aq)} + \frac{1}{2}O_{2(g)} + 2e^{-} \rightarrow ZnO_{(s)} \quad (6)$$

Zinc, in oxy-anions presence, has characteristic behaviors of anodic inhibitors, favoring a direct formation of metal oxides and oxides coming right from reduction of its own inhibitor. The displacement of potential for more positive values is result of oxides stability formed in a slightly alkaline solution. In solution containing chromate, the ZnO and Cr_2O_3 formation can be considered as a result of corrosion inhibiting effect on zinc, according to Equation (7). Similarly, in molybdate presence, Equation (8), there is formation of ZnO and MoO_2 or MoO_3 [6].

$$3Zn + 2CrO_4^{2-} + 2H_2O \rightarrow 3ZnO + Cr_2O_3 + 4OH^- \quad (7)$$

$$Zn + MoO_4^{2-} + H_2O \rightarrow ZnO + MoO_2 + 2OH^- \quad (8)$$

The oxy-anions tungstate and molybdate are very effective corrosion inhibitors in localized corrosion inhibition of steels their alloys in media containing chloride. The tungstate action way is similar to oxy-anion molybdate. Tungsten can form a large number of complex and, in slightly acidic solutions, they form insoluble oxides such as WO_2, W_2O_5 and WO_3. In presence of alkaline solutions, they have a slight tendency to decompose water, releasing hydrogen, being dissolved in hexavalent state as tungstate ions (WO_4^{2-}). The WO_4^{2-} anion has properties similar to a buffer, consuming H^+ ions, which are produced in hydrolysis of metal cations Equation (9) suppressing the local acidification. In addition, polymeric species of tungstate are formed with metal cations, Equation 10, and can be adsorbed on metal surface, repairing defects and pits, producing a stable passive layer [7].

$$WO_4^{-2} + H^+ \rightleftarrows HWO_4^- \quad (9)$$

$$HWO_4^- + H^+ + 2H_2O \rightleftarrows W(OH)_6 \quad (10)$$

3.2. Potentiodynamic Polarization Curves

From electrochemical studies, potentiodynamic polarization curves were carried out with electrodeposits of Zn and Zn-Fe-Co before and after the blue chromatization, at a scan rate of about 0.333 $mV \cdot s^{-1}$ (20 $mV \cdot min^{-1}$), starting in a –0.30 V potential below the corrosion potential. In **Figure 2**, curves were obtained in aerated solution of 3.5% NaCl, pH 8.2 and it was observed in cathodic region both for the electrogalvanised and the alloy, with and without passivation, a limit current which indicates a diffusional control, i.e., the procedure is controlled by the oxygen diffusion due to oxygen reduction reaction on metal.

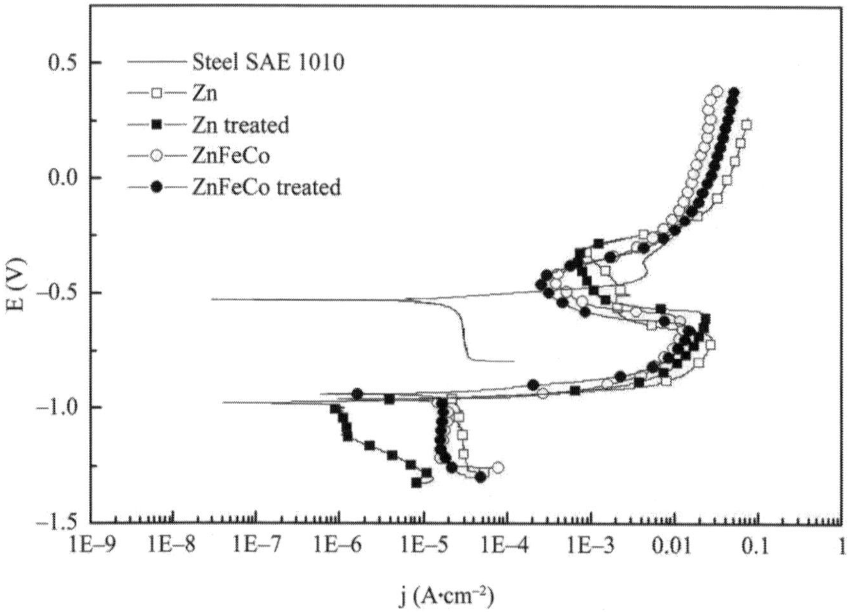

Figure 2. Potentiodynamic polarization curves in 3.5% NaCl, pH 8.2, for steel and electrodeposits with and without treatment.

For treatment-free samples, Zn and Zn-Fe-Co showed current densities of about 1.8×10^{-5} A·cm^{-2}and 2.5×10^{-5} A·cm^{-2}, respectively. After zinc chromatization, a clear change in cathodic region occurs, where initially a current decrease appears, attributed to the barrier effect carried out by CrO_4^{2-} layer [8] and also presence of limit current of oxygen reduction, which in this case is much lesser, of about 1.1×10^{-6} A·cm^{-2}. For the recorded curves, it is verified that chromatization effect does not affect corrosion potential, because for these coatings the potential value is approximately –1.0 V, that is, they remain constant. Therefore, it is noted that although chromate layer acts effectively in corrosion current density reduction, it has little effect on localized corrosion resistance.

After corrosion potential in anodic region, it is noticed an active zinc dissolution reaching a current density of about 20 mA·cm^{-2}. For potential values close to –0.4 V, there is a current decrease, which is associated with zinc removal and consequent steel exposure; above this potential only substrate dissolution occurs.

Figure 3(a) shows that when oxy-anions are added to chloride solution, it is observed a change in reduction process mechanism, probably due to reduction of these anions on electrode [9]. These anions addition shifts the electrogalvanised corrosion potential and Zn-Fe-Co alloy to more positive values, indicating an increase on the localized corrosion resistance of metal in such media, also in all cases a cathodic current decrease occurs. Following results analysis, compared to other anions, chromate is the inhibitor that provides the most zinc protection. An analysis of corrosion potential values (E_{corr}) and corrosion current density (j_{corr}) also shows that chromate presents a better performance **Table 1**.

In aerate medium, the two samples have shown limit current of oxygen reduction and results have not revealed a passive behavior. When the systems were deaerated with nitrogen gas bubbling at room temperature of about 25°C, the experiments conducted in 3.5% NaCl exhibited, in cathodic region, a typical behavior of activation control and in anode region a slight passivation tendency for Zn and Zn-Fe-Co samples. Corrosion potentials obtained in aerated medium were approximately –0.97 V and in deaerated medium –1.17 V, **Figure 3**(b).

3.3. Cyclic Voltammetry

At a speed of 33.3 mV·s^{-1}, the cyclic voltammetry tests have been carried out, starting a scan at –1.5 V towards anodic potentials with potential reversal in region where material dissolution is observed.

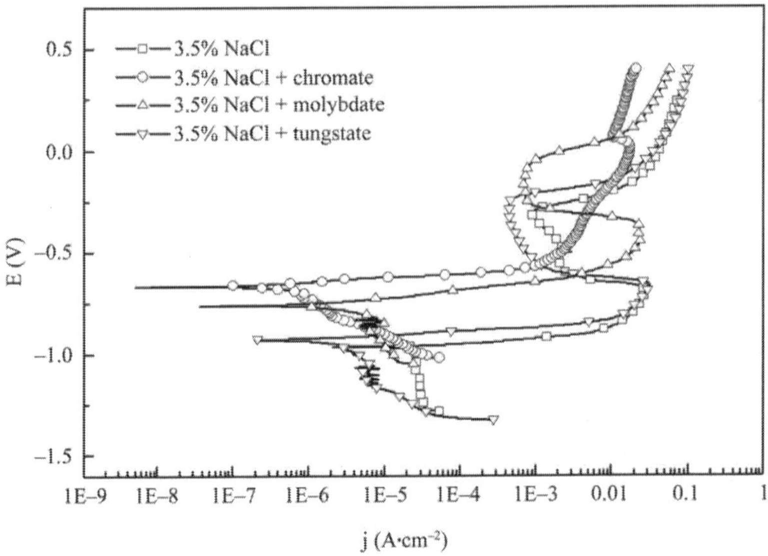

(a)

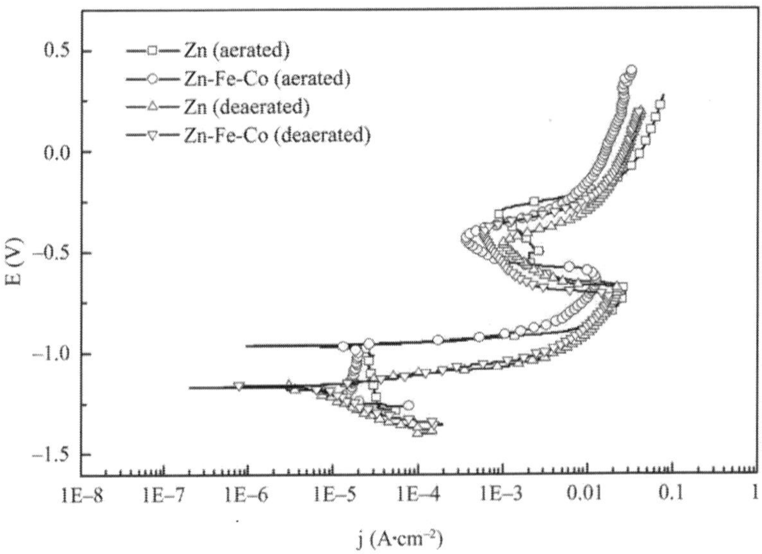

(b)

Figure 3. Potentiodynamic polarization curves for electrodeposits: (a) pure Zn in a 3.5% NaCl solution, pH 8.2; and in chloride with addition of CrO_4^{2-}, MoO_4^{2-} or WO_4^{2-} 0.05 mol·L^{-1}; (b) for Zn and Zn-Fe-Co in aerated and deaerated chloride solution.

Table 1. Corrosion potential values (E_{corr}) and corrosion current density (j_{corr}) for pure Zn and Zn-Fe-Co, in 3.5% NaCl solution and with addition of CrO_4^{2-}, MoO_4^{2-} or WO_4^{2-} 0.05 M.

Zn	NaCl	NaCl + chromate	NaCl + molybdate	NaCl + tungstate
E_{corr}/V	−0.97	−0.70	−0.76	−0.94
j_{corr}/A·cm^{-2}	2.8×10^{-5}	5.8×10^{-7}	3.6×10^{-6}	2.8×10^{-6}
Zn-Fe-Co	NaCl	NaCl + chromate	NaCl + molybdate	NaCl + tungstate
E_{corr}/V	−0.94	−0.69	−0.63	−0.77
j_{corr}/A·cm^{-2}	2.4×10^{-5}	6.6×10^{-7}	5.4×10^{-6}	6.9×10^{-6}

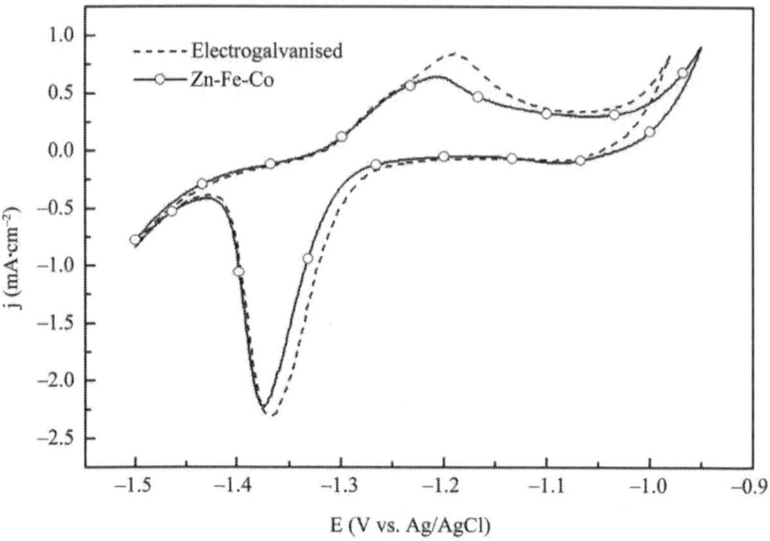

(a)

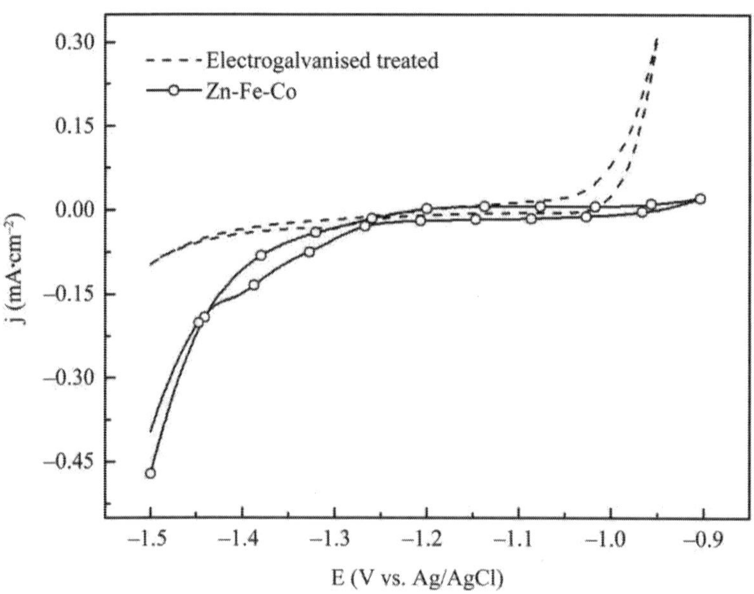

(b)

Figure 4. Cyclic voltammograms for electrodeposits of Zn and Zn-Fe-Co on steel in 3.5% NaCl solution, pH 8.2: (a) without treatment and (b) with treatment.

Cyclic voltammograms for electrodeposits of Zn and Zn-Fe-Co are shown in **Figure 4**(a). Anodic process is characterized by presence of an oxidation current peak at about –1.2 V, attributed to an oxide film or zinc hydroxide formation on electrode. Then it is observed the metal dissolution, which occurs beginning at –1.0 V. In this potential region, it is possible to verify that zinc coating has less stability than the Zn-Fe-Co electrodeposit, whose dissolution occurs at more anodic potentials. After the potential scan inversion, it is observed, towards negative potentials, a reduction current peak localized at –1.37 V, associated with reduction of the film formed in oxidation process. This reduction process becomes more complex and its load increases when the anodic reversal potential reaches higher values, where the present concentration of soluble zinc species becomes elevated. For chromatized electrodeposits **Figure 4**(b), it is observed that occurs a significant charge reduction and this is related to oxidation and reduction processes in zinc oxide/hydroxide formation region, which indicates the efficiency of the applied process.

4. CONCLUSIONS

From the electrochemical tests carried out, it can be concluded that the steel subjected to electrodeposition process and then to chromatization treatment has shown to be more corrosion resistant than the passivation-free electrodeposits. The corrosion potential remained almost unchanged, whereas a significant decrease at corrosion current density has occurred for the pure Zn treated sample. The steel substrate exhibits a corrosion potential of approximately –0.5 V, which shows a big difference among the electrodeposits potentials in study indicating that they protect steel by acting as a sacrificial metal.

By adding oxy-anions to 3.5% NaCl solution it has resulted in a significant reduction of corrosion current density, and in a displacement of corrosion potential to more noble values.

Both for Zn and Zn Fe Co, deaerating NaCl solution has resulted in the elimination of limit current of oxygen reduction, indicating a mechanism change of the cathodic process.

5. ACKNOWLEDGEMENTS

Authors thank to COOKSON ELECTRONICS BRAZIL LTDA and FUNDUNESP (process 01258/08-DFP).

REFERENCES

1 R. Favaron and L. M. Aleixo, "Voltammetric determination of Zn (II) in Zn-Fe Alloy Electroplating Baths Using Square-Wave Voltammetry," Journal of the Brazilian Chemical Society, Vol. 12, No. 2, 2001, pp. 173-179.

2 A. Brenner, "Electrodeposition of Alloys," Academic Press, New York, 1963.

3 P. L. Neto, A. O. Souza, M. I. C. Nogueira and R. P. Colares, "Estudos da corrosão de eletrodepósitos de Zn-Ni, Zn-Fe e Zn-Co em meio de cloreto e de sulfato," Coteq, Salvador, 2002.

4 Z. Wu, L. Fedrizzi and P. L. Bonora, "Electrochemical studies of Zinc-Nickel Codeposition in Chloride Baths," Surface and Coatings Technology, Vol. 85, No. 3, 1996, pp. 170-174.

5 E. Tada, S. Satoh and H. Kaneko, "The Spatial Distribution of Zn^{2+} during Galvanic Corrosion of a Zn/steel couple," Electrochimica Acta, Vol. 49, No. 14, 2004, pp. 2279- 2285.

6 C. B. Breslin, G. Treacy and W. M. Carrol, "Studies on the passivation of aluminium in chromate and molybdate solutions," Corrosion Science, Vol. 36, No. 7, 1994, pp. 1143-1154.

7 D. S. Azambuja, E. M. A. Martini and I. L. Muller, "Corrosion behaviour of iron and AISI 304 Stainless Steel in Tungstate Aqueous Solutions Containing Chloride," Journal of the Brazilian Chemical Society, Vol. 14, No. 4, 2003, pp. 570-576.

8 R. Fratesi and G. Roventi, "Corrosion resistance of ZnNi Alloy Coatings in Industrial Prodution," Surface and Coatings Technology, Vol. 82, No. 1-2, 1996, pp. 158- 164.

9 G. D. Wilcox and D. R. Gabe, "Passivation Studies Using Group VIA Anions. Part 5: Cathodic treatment of zinc," Britsh Corrosion Journal, Vol. 22, No. 4, 1987, pp. 254- 258.

CHAPTER 4

Trend of Nitriding on Chromium-Molybdenum Steel via Low Temperature Screen Plasma Technology

Kuk-Hyun Yeo[1], Sang-Gweon Kim[1*], Jae-Hoon Lee[1], Jung-Hyun Kong[2], Masahiro Okumiya[2]

[1]Heat Treatment Technology R&BD Group, Korea Institute of Industrial Technology, Incheon, Korea

[2]Toyota Technological Institute, Nagoya, Japan

ABSTRACT

Low temperature screen plasma technology, a high plasma density, through using a low energy supply, shows excellent effects on a low alloy chromium-molybdenum steel for plastic molds because it does not show a compound layer and a high surface hardness without a deterioration in matrix hardness. For interest about hardening depth, both the screen plasma nitriding and plasma nitro-carburizing processes were tested including nitrogen, hydrogen and a methane mixed gas environmental at 653 K, 713 K. The optical emission spectroscopy (OES) has been analyzed during screen plasma nitriding (SPN) and a nitro-carburizing process (SPNC) was proceeded at 713 K and the same pressure. I find it difficult to dissociate nitrogen molecules perfectly with neutral nitrogen atoms via the DC-plasma nitriding process. Therefore, the SPN and SPNC process have shown a high density of plasma species even though low temperature plasma conditions have a high peak intensity of H_β and H_γ in the results of the analysis by OES. The hardness value was measured with the micro-Vickers hardness tester after the SPN, SPNC process and the chemical composition of nitriding layers were traced by GDOES. The screen nitriding layer via the screen plasma technology has shown excellent properties with a thickness depth of about 850 ~ 900 HV without the deterioration of matrix hardness value.

KEYWORDS

Screen Plasma Technology, Plasma Nitro-Carburizing, Optical Emission Spectroscopy, Without Compound Layer, Chromium-Molybdenum Steel

1. INTRODUCTION

In recognition of its eco-friendliness, the demand for the plasma nitriding process has exploded in the area of surface hardening technology for energy efficiency, low distortion and high performance in the past few decades. And the screen plasma technology is used as a method of making low energy assumption for high density plasma and high efficiency in the plasma nitriding process [1] -[4] . At last, advances in the screen plasma technology like active screen plasma nitriding or ATONA process have made it possible for low temperature hardening to low-alloying steel with a simple design and low energy assumption [5] . We took it for granted that the screen plasma process was issued with a high-density plasma species as a neutral nitrogen atom and radical ammonia molecule, example NH_x, to supply floating power by remote electrodes [6] -[9] .

Recently, the influence of screen plasma technologies has been applied to the development of a hardening process breaking through the barrier [1] [10] . To understand the principle of resource generation, most researchers were interested in plasma species models to describe the stages of plasma species generation and the movement in chamber including the remote plasma process as well as how it differs from the mass transfer mechanism in the plasma nitriding process and active screen nitriding process, particularly the example sputtering model and the role of dominant plasma species, etc. [11] -[15] .

On the one hand, the commonly used low alloyed mold steel was naturally treated using a quenching-andtempering hardening process at low temperature, and it used the gas oxi-nitriding process for wear test atmosphere. In this research, it will thus be seen that the low-element alloying chromium-molybdenum steel has been approached to enhanced performance without a compound layer and matrix hardness deterioration by the SPN and SPNC process and presented our diagnostics results with some parameters in applying a SPN and SPNC technique to AISI4140 which satisfies the above-mentioned requirements.

In particular, the screen plasma process in a mixed gases atmosphere holds the key in the plasma reaction of the nitriding, nitro-carburizing, and carburizing process. However, if the activated gas species will be analyzed by optical spectroscopic equipment, understanding the nitrogen diffusion mechanism would be deeper and could attribute to optimizing the process condition.

In our research, we mainly investigated the behavior of the neutral ionized gas species with variations of temperature and the composition ratio of gas species (N_2, H_2, CH_4) by Optical Emission Spectroscopy (OES) during the screen plasma process in the chamber which has a double-folded screen. From the surface analysis by OM and SEM, and Glow-Discharge Optical Emission Spectroscopy (GDOES), we could find out how to dissociate to a neutral and an ionic nitrogen and to contribute to diffuse in matrix material. And it has shown a high concentration of nitrogen with the same hardness profiles without a compound layer and the low hardness deterioration of Quenching-and-Tempering treated AISI4140.

2. EXPERIMENTAL DETAILS

2.1. Materials

The used AISI4140 of chemical composition (wt%) is showed as: 0.38% - 0.43% C; 0.15% - 0.35% Si; 0.6% - 0.9% Mn; 0.03% P; 0.03% S; 0.9%

- 1.2% Cr; 0.15% - 0.3% Mo; Fe balance (Table 1). The samples to be treated were machined in the form of 30 mm diameter and 15 mm thick discs. The chromium-molybdenum steel sample had 1123 K quenching 1 hour, the 473 K tempering preserving time was about 2 hours. The sample surfaces were mechanically polished sequentially by 220, 400, 600, 800, 1500 and 2000 grit wet SiC emery paper, finely polished with 0.03 μm alumina suspension, further degreased ultrasonically in acetone, cleaned in ethylalcohol, and dried in air blowing before being placed in the screen plasma process chamber.

2.2. Experimental Apparatus

The SPN and SPNC process was carried out using by plasma equipment in Figure 1. It contains a cylindrical stainless-steel vacuum chamber with the dimensions of 1300 mm diameter and a length of 2600 mm. The samples were placed in a double screen set-up where it was directly put on the electrically insulated stage (cathode) of the plasma nitriding unit (the chamber wall served as anode). The double screen was a mesh cylinder with a diameter of 900 mm and a height of 1800 mm, and it was made of 1.0 mm. The double-folded screen insulated with a power supply has been laid inside the chamber and a sample of SPN or SPNC was laid on a Jig with a floating power supply.

Table 1. Chemical composition of AISI4140.

AISI4140	C	Si	Mn	P	S	C	Mo	Fe
Concentration (wt%)	0.38 - 0.43	0.15 - 0.35	0.60 - 0.90	~0.03	~0.03	0.90 - 1.20	0.15 - 030	Bal.

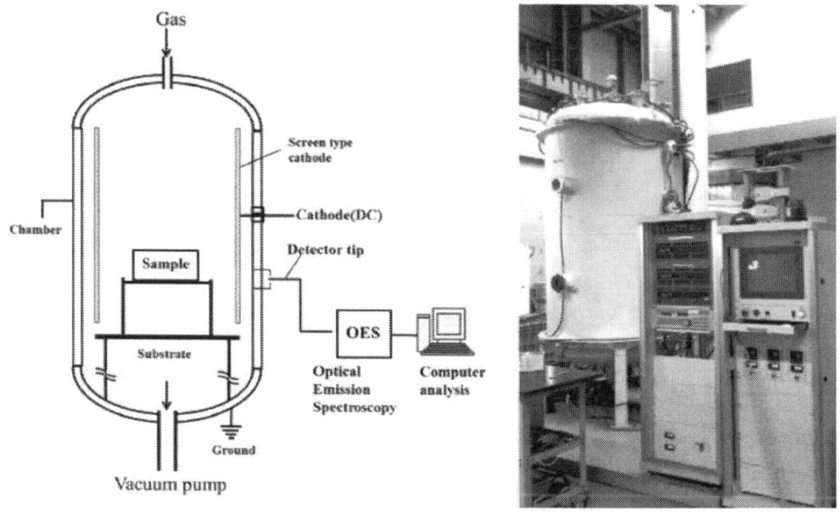

Figure 1. Experimental concept and screen plasma equipment.

2.3. Screen Plasma Nitriding (SPN) and Screen Plasma Nitro-Carburizing (SPNC) Processing

Table 2 shows the details of each plasma treatment with specimen symbols. After an evacuation of up to 1.33 Pa, the process pressure of the mixed gas for each specimen was adjusted to 20 Pa.

Figure 2 has showed an experimental procedure of a gas flow rate and the process temperature included by soaking the steps of uniformity for about two hours.

We mainly investigated the behavior of the plasma nitriding and nitro-carburizing process at a low temperature with variations of time conditions and a continuous ratio of gas composition (N_2, H_2, CH_4).

Table 2. Processing parameters for SPN and SPNC.

Process	Temperature (K)	Time (hr)	Pressure (Pa)	Gas Composition (sccm) (N_2:H_2:CH_4)	Power (kW)
Screen Plasma Nitriding (SPN)	653,713	4, 10, 20	20	80:240:0	~35
Screen Plasma Nitro-Carburizing (SPNC)	653,713	4, 10, 20	20	80:240:5	~40

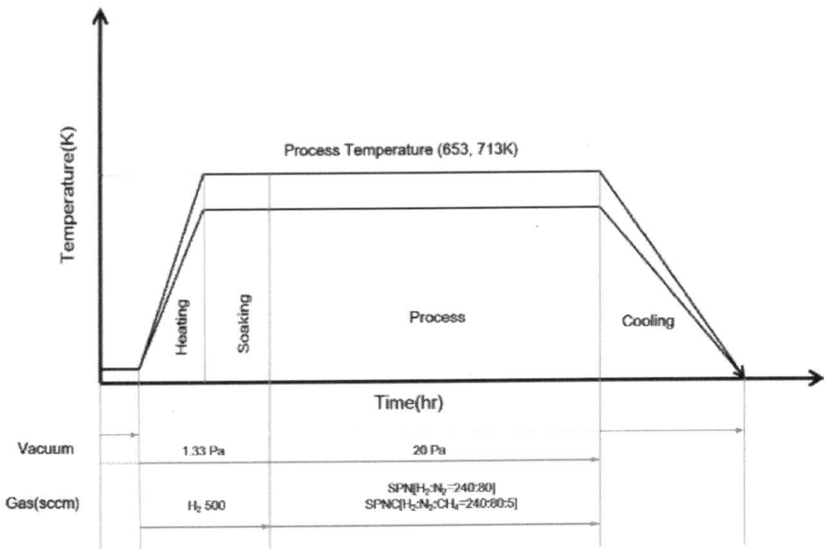

Figure 2. The experimental cycle of SPN and SPNC.

The SPN specimen was screen plasma treated at 653 K and 715 K with a mixture of 30% nitrogen gas and 70% hydrogen gas for 4, 10, 20 hr and The SPNC specimen was treated using a screen plasma nitro-carburizing process for 4, 10, 20 hr at 653 K and 713 K with a mixture of nitrogen, hydrogen, and methane gas.

2.4. Analytical Methods

We have traced plasma species in more detail by optical emission spectroscopy (OES, AVANTE AvaSpec-3648) during the screen plasma process. The microstructure analysis was carried out using both optical microscopy (OM: Olympus-BX51TRF) and scanning electron microscopy (SEM: JEOL FE-SEM/EDS JSM-7001F). It was analyzed by glow discharge emission spectrometry (GD-OES: GDA750/SPECTRUMA ANALYTIK GmbH) a composition of hardening layers as functions of time and temperature. The surface hardness and the case depth profiles were determined using a micro-vickers hardness tester (Future Tech FM-7).

3. RESULTS AND DISCUSSION

OES analysis is a powerful diagnostic tool for the identification of atomic and molecular species present in the plasma used for surface treatment. We have observed especially radiative transitions of prominent band systems of nitrogen, hydrogen, methane and sputtered Fe as depicted in Table 3 The relative emission intensities vs. the nitrogen concentrations are shown in Figure 8. After a low-pressure discharge (0.1 torr) in N_2-H_2 plasma, in which the screen, and the gas mixture are introduced into the vacuum chamber and the potential, is applied to the cathode. The spectral line intensities are taken when they have reached a stable value. The most intense spectral lines are measured from the excited ions at a wavelength of N_2^* (391.4/427.8 nm) and H_γ (434.1 nm).

Table 3. Prominent spectral lines of N_2-H_2-(CH_4) plasma [6] -[9] [15] .

Species	Transition Array	Spectral Line (nm)	Species	Transition Array	Spectral Line (nm)
N_2	$B^3\Pi_g$-$A^3\Sigma_u$	580.4 662.3 775.3	H	Balmer series	656.3 (H_α) 486.1 (H_β) 434.1 (H_γ)
N_2	$C^3\Pi_u$-$B^3\Pi_g$	380.5 337.1	CN	$B^2\Sigma$-$A^2\Pi$	419.8 423.9 429.2
N_2^+	$B^2\Sigma_u^+$-$X^2\Sigma_g^+$	391.4 427.8	CN	$A^2\Pi$-$X^2\Sigma$	585.8 647.8 663.1
NH	$A^3\Pi$-$X^3\Sigma^-$	336	N^+	2p3s-2p3p	568.0
NH_3	Schuster's emission band	528.2 563.5 567.0 649.7 766.6	C_2	$A^3\Pi_g$-$x'^3\Pi_u$	410.2 468.0 563.5 589.9 790.7 810.8
N	$2p^23s(^4P)$-$2p^23p(^4S^0)$	744.2 746.8	Fe	Z5F-a^5D	372.0 561.6
N	Spectra handbook [15]	399.5 463.0 500.5 575.2	CH	$B^2\Sigma$-$X^2\Pi$	387.1 388.9 432.4
O_2	$A3\Sigma_u^+$-$X^3\Sigma_g^-$	337.0 345.3			

Figure 3 and Figure 4, intensities of the two measured Balmer lines H and various plasma species of hydrogen atoms were shown at a high intensity. The analysis of plasma species generated under the SPN and SPNC conditions detected various nitrogen active species.

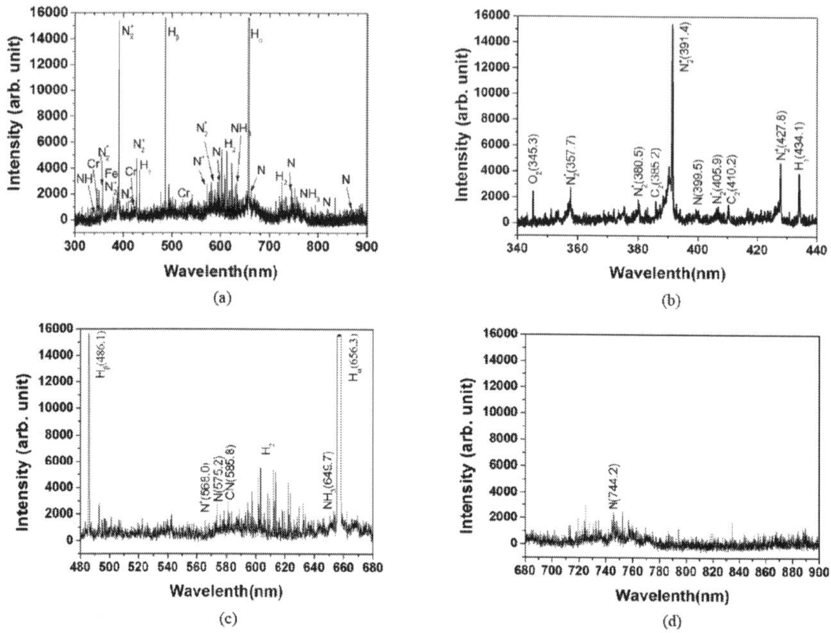

Figure 3. Results of optical emission spectroscopy analysis for SPN. [713 K, N_2:H_2 = 80:240 (sccm)]. (a) Full range; (b) From 340 to 440 (nm); (c) From 480 to 680 (nm); (d) From 680 to 900 (nm).

Figure 3(a) and Figure 4(a) confirm the generation of species with very intense and high peak, and the fact that H_γ which is only in high energy plasma, was observed in relative high intensity in addition to H_α and H_β shows that the process can create high energy although it is in the very low energy state. As shown in Figure 3(b), Figure 3(c), and Figure 3(d), the N and N+ peaks in the atomic form were generated in significant intensity. On the other hand, (Figure 4(a)), the addition of small quantities of methane (less than 5 sccm) leads to a significant increase in intensity. The increase in emission intensity of nitrogen with the addition of small quantities of methane can be explained due to an increase N_2^* (427.8 nm) in the secondary electron emission coefficient.

Figure 4(b) and Figure 4(c) show CH and CN peaks observed in various wavelengths although CH (388.9 nm) is observed only in the SPNC process. Large amount of high intensity neutral nitrogen species along with $N_2^{\bullet}$ and N_2^{+} species, which are the active nitrogen species, and NH_x plasma species were observed in both SPN and SPNC processes.

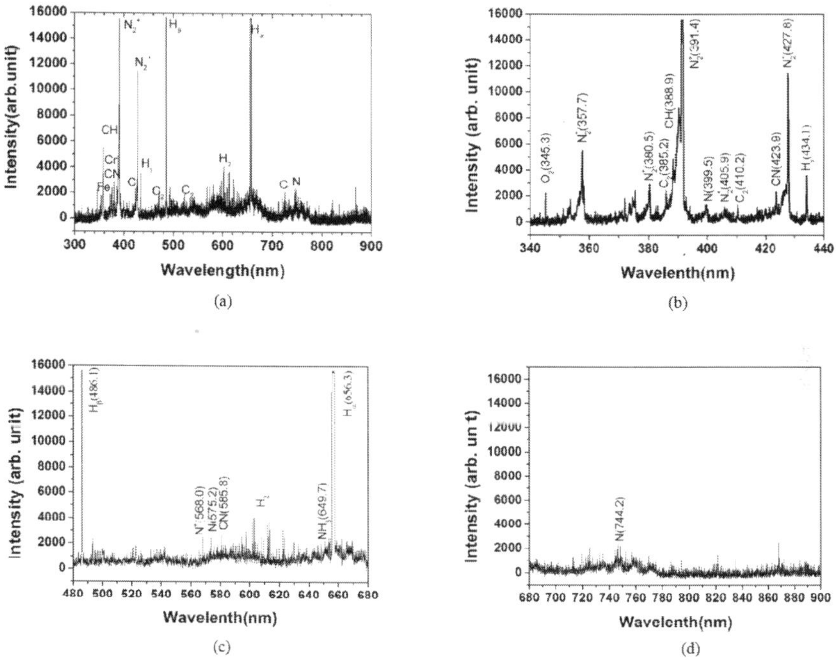

Figure 4. Results of optical emission spectroscopy analysis for SPN. [713 K, N_2:H_2:CH_4 = 80:240:5 (sccm)]. (a) Full range; (b) From 340 to 440 (nm); (c) From 480 to 680(nm); (d) From 680 to 900 (nm).

In the SPNC process particularly, increasing intensities of plasma generating species were observed by injection mixing of small amount of methane gas, Although C_2 was observed in both SPN and SPNC processes, the CH plasma was observed only in the SPNC process.

It is because C_2 remaining on the reactor wall was sputtered as the effect of plasma through the several processes while the CH plasma specifies were observed only in the SPNC process.

Figure 5 depicts the result of surface condition observation by SEM in SPN. It shows coating of spherical grains after sputtering on screens in sizes of dozens nm to hundreds nm. As indicated by the result of OES observation shown in Figure 3(a) and Figure 4(a), Fe and Cr are sputtered on the surface of the double-folded screen electrode, and the sputtered Fe and Cr metals react with high density ionized species as they move inside the plasma then form the spherical grains on the surface of the nitriding layer.

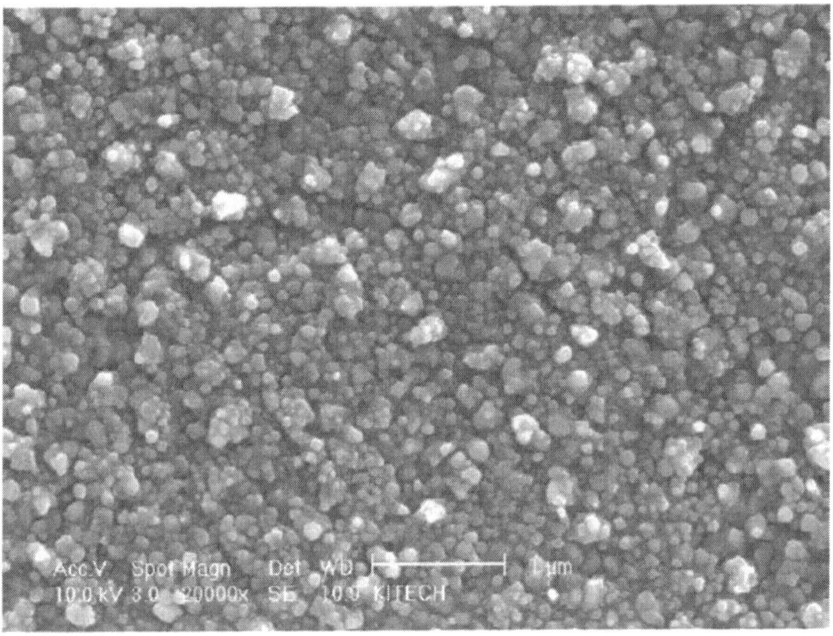

Figure 5. SEM surface image of SPN treated specimen.

After the specimen, both screen plasma nitriding and nitro-carburizing had been cut and mounted for microscopic analysis, at Figure 6 have shown images of SPN with a treating time for (a) 4 hr, (b) 10 hr, (c) 20 hr at 653 K. It has clearly shown a tempered martensite structure without a compound layer on the low-temperature SPN conditions with time of Figure 7(a), Figure 7(b), and Figure 7(c). Though according to the process conditions, the screen nitriding process can cause a chemical compound layer (also known as white layer) to occur on the surface of the steel.

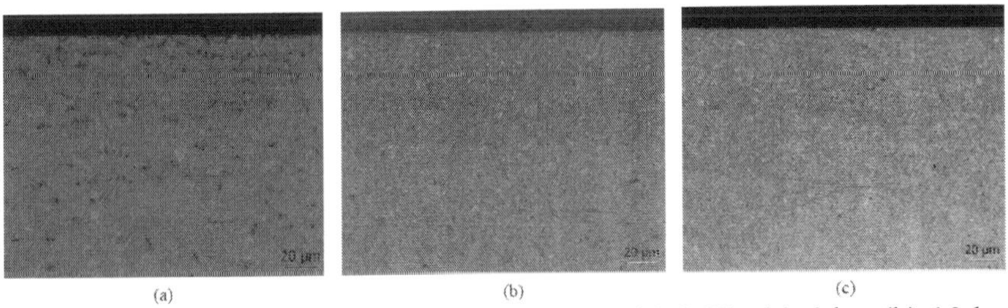

Figure 6. Structure at different hours of SPN (653 K). (a) 4 hr; (b) 10 hr; (c) 20 hr.

Figure 7 shows the pictures of specimen SPN treated at 713 K. (b) and (c) confirm the formation of nitrogen compound layer of 1 μm or less on the surface for 10 hours or longer.

Figure 8 shows the pictures of micro structures at each hour of SPNC treatment at 653 K as well as the cross section in the SPNC process.

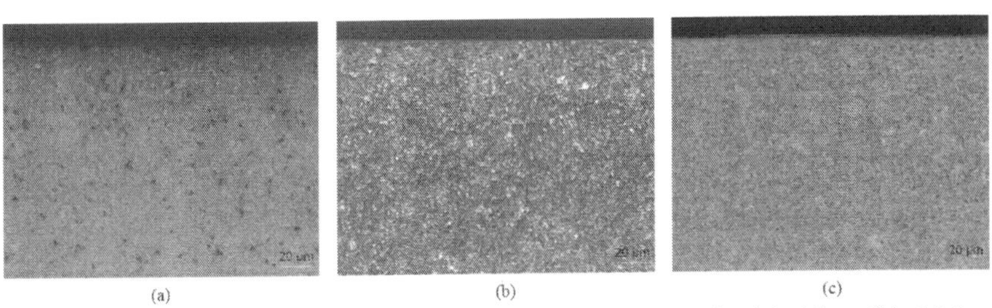

Figure 7. Structure at different hours of SPN (713 K). (a) 4 hr; (b) 10 hr; (c) 20 hr.

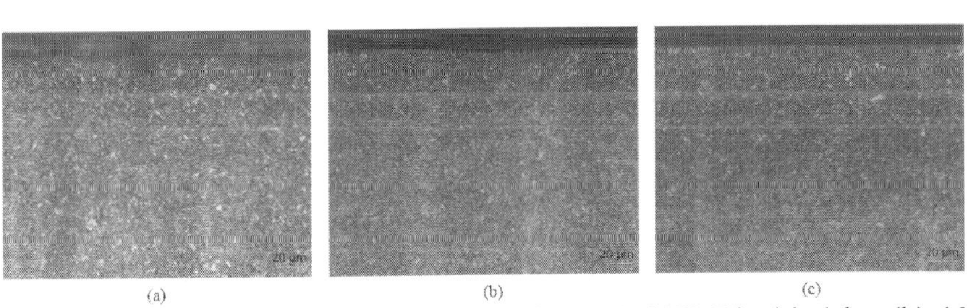

Figure 8. Structure at different hours of SPNC (653 K). (a) 4 hr; (b) 10 hr; (c) 20 hr.

Figure 9 shows the pictures of organization under the SPNC 713 K condition. It indicates the hardening depth and degree of etching by expansion of nitrogen are not much different.

Figure 10 shows the micro-hardness profiles an obtained of the SPN as functions of temperature at 653 K and 713 K. Somewhat low measurement of internal hardness at 653 K for 10 hours and at 713 K for 4 hours in Figure 10 can be attributed to somewhat low measurement of hardness of material inside due to the uneven heat treatment during the Q-T process because of the properties of AISI4140 material and overall low hardness of test samples. Still, it shows the trend of 30% - 40% hardness increase over the original material.

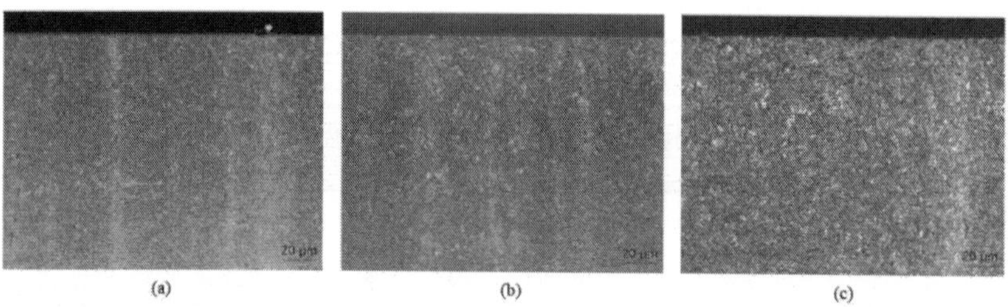

Figure 9. Structure at different hours of SPNC (713 K). (a) 4 hr; (b) 10 hr; (c) 20 hr.

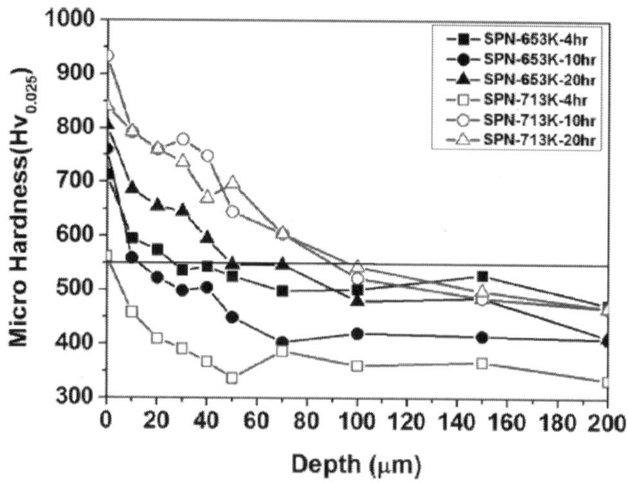

Figure 10. Hardness profile at different hours of SPN process.

Figure 11 depicts the hardness profile of SPNC. It shows the effective hardened layer of hardness of 550 Hv or more and the depth of 25 μm for 4 hours and 90 μm for 10 hours and 20 hours at 713 K. It confirms hardening of surface only without hardness deterioration of core. The surface hardness was 700 Hv and effective hardened layer depth was 30 μm. The surface hardness was 760 Hv and depth was 7 μm in the case of 10 hours, and they were 770 Hv and 100 μm in the case of 20 hours.

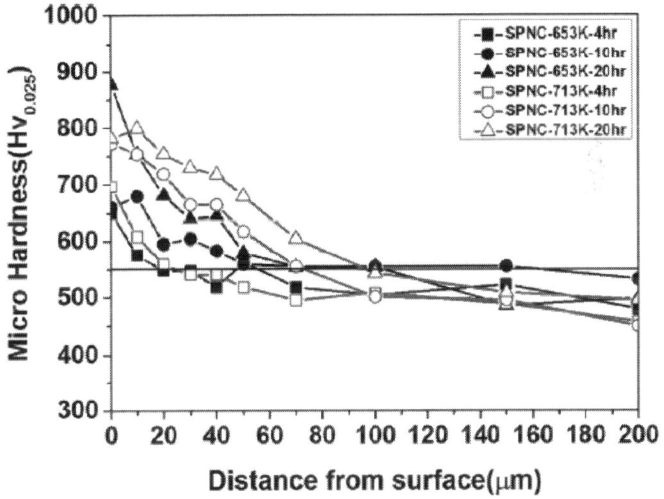

Figure 11. Hardness profile at different hours of SPNC process.

It was also important to achieve a surface structure with deep diffusion and even without a compound layer and which is characterized by a high nitrided diffusion layer occurring at a depth of about 90 μm with the SPN process and about 100 μm with the PNC process. The analysis has revealed that these granular diffusions consist of a nitrided layer.

Figure 12 depicts the result of nitrogen and carbon chemical composition analysis by GDOES for SPN treatment of (a) 4 hr, (b) 10 hr and (c) 20 hr at 653 K.

Figure 12(a) shows the maximum nitrogen concentration of 4.8 (at%) and carbon concentration of 7 (at%) for 4 hours of SPN treatment; (b) shows the maximum nitrogen concentration of 9.4 (at%) and carbon concentration of 7.5 (at%) for 10 hours of SPN treatment; and (c) shows the maximum nitrogen concentration of 11.7 (at%) and carbon concentration of 6.5 (at%) for 20 hours of SPN treatment.

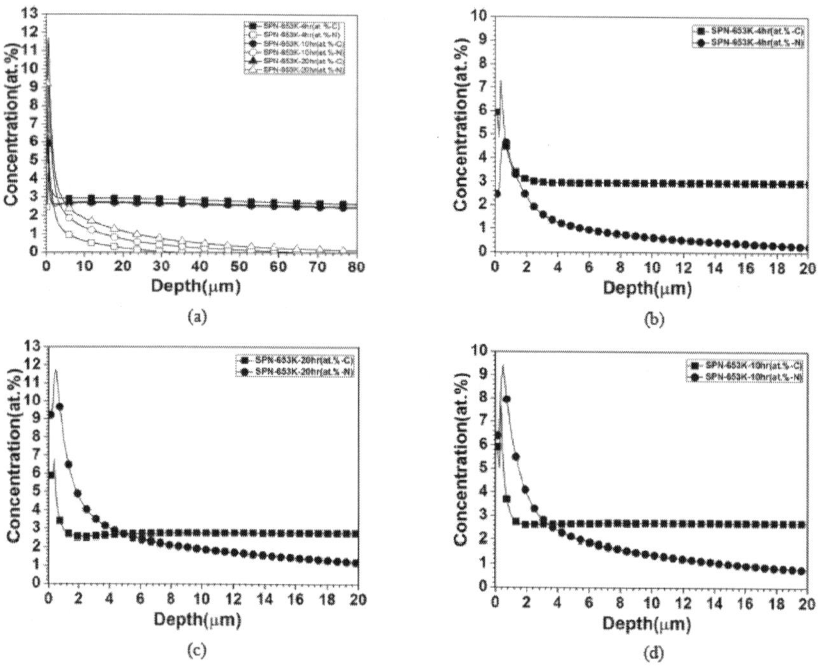

Figure 12. It shows the GDOES depth profile of simultaneous SPN at 653 K. (a) Full range; (b) 4 hr; (c) 10 hr; (d) 20 hr.

Figure 13 depicts the result of chemical composition analysis of SPN treated at 713 K. It shows the very high nitrogen and carbon concentrations of (a) 34.1 (at%) and 6 (at%), respectively for 4 hours, (b) 47 (at%) and 10 (at%), respectively for 10 hours, and (c) 32.4 (at%) and 13 (at%), respectively for 20 hours.

Comparison with the hardness profile of the case of SPN process at 653 K and 713 K shown inFigure 10 indicates that the change of nitrogen concentration + carbon concentration and surface hardness almost coincide. Figure 12 and Figure 13 depict total nitrogen and carbon contents and show the hardness increase to significant depth even with small amount of nitrogen and carbon.

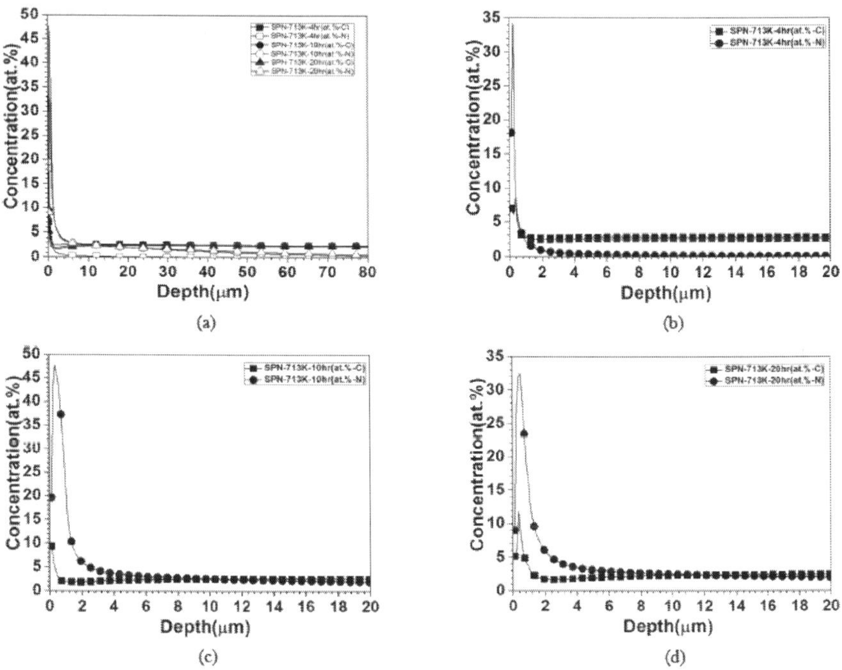

Figure 13. It shows the GDOES depth profile of simultaneous SPN at 713 K. (a) Full range; (b) 4 hr; (c) 10 hr; (d) 20 hr.

Figure 14 depicts the result of SPNC 653 K. It shows the maximum nitrogen concentration and carbon concentration of (a) 7.5 (at%) and 6 (at%), respectively for 4 hours, (b) 5.1 (at%) and 8 (at%), respectively for 10 hours, and (c) 16.8 (at%) and 6 (at%), respectively for 20 hours. Figure 15 depicts the result of SPNC 713 K. It shows the nitrogen concentration and carbon concentration at different hours of treatment (a) 38.9 (at%) and 12.5 (at%), respectively for 4 hours, (b) 36.4 (at%) and 17 (at%), respectively for 10 hours, and (c) 44.8 (at%) and 18 (at%), respectively for 20 hours. Compared to Figure 13 which depicts the result of SPN process at the temperature, they show very high nitrogen and carbon concentrations.

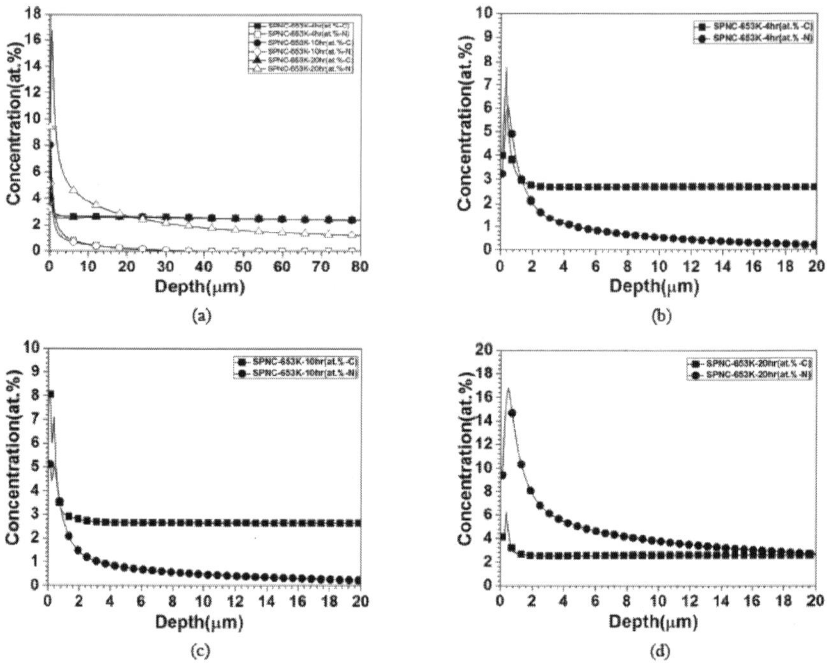

Figure 14. It shows the GDOES depth profile of simultaneous SPNC at 653 K. (a) Full range; (b) 4 hr; (c) 10 hr; (d) 20 hr.

The results of SPNC hardened layer depths in Figure 11 and SPN hardened layer depths in Figure 10 SPNC indicate that the change of surface hardness is more dominated by nitrogen concentration than the carbon concentration.

However, in the SPNC process where the carbon is additionally supplied, the hardness of the lower level layer did not deteriorate greatly and thus the deeper effective hardened layer was obtained.

As shown by Figure 12 to Figure 14, GDOES result indicates that the 1 micron deep layer formed by the grain shape objects formed by sputtering shown in Figure 5, is the nitrides with high nitrogen concentration.

The GDOES result shows that the nitrogen and carbon, which increase the hardening depth, can maintain the high hardness even with a low concentration of a few %.

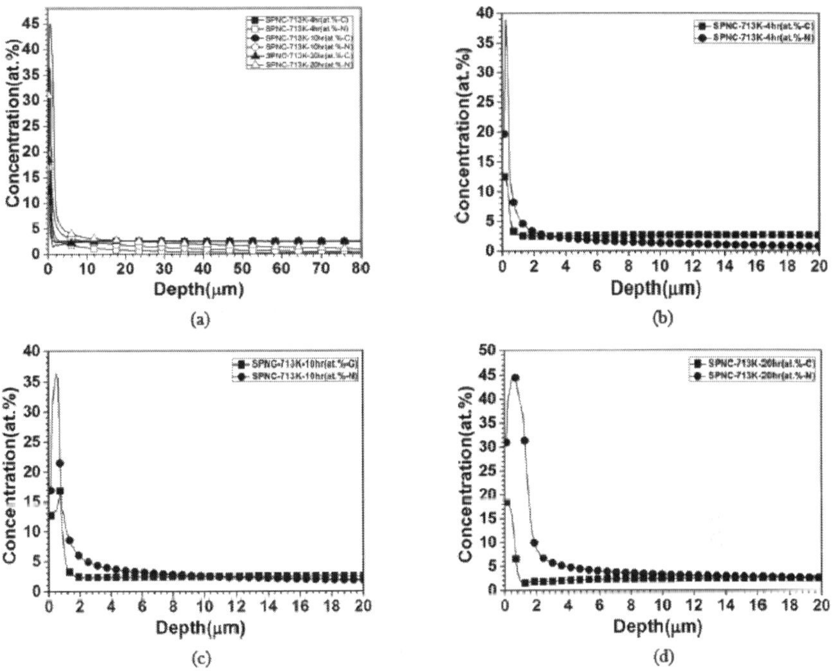

Figure 15. It shows the GDOES depth profile of simultaneous SPNC at 713 K. (a) Full range; (b) 4 hr; (c) 10 hr; (d) 20 hr.

4. CONCLUSIONS

1) Low temperature tempered chromium-molybdenum steel was treated in screen plasma nitriding (SPN) and nitro-carburizing (SPNC) process at temperatures of 653 K and 713 K for 4 hr, 10 hr and 20 hr.

2) The analysis of plasma species generated under the SPN and SPNC conditions indicated various nitrogen active species in both SPN and SPNC. Particularly the fact that H_γ which is only in high energy plasma, was observed in relative high intensity in addition to the fact that H_α and H_β shows that the process can create high energy although it is in the very low energy state.

3) Large amount of high intensity neutral nitrogen species along with N_2^* and N_2^+ species, which are the active nitrogen species, and NH_x plasma species were observed in both SPN and SPNC processes.

4) SNPC is particularly affected by the mixed methane gas to increase the intensity of overall plasma species. That is proven by observation of C_2 in both SPN and SPNC processes while CH plasma species were observed only in the SPNC process. As the process is executed several times, C_2 remaining in the reactor appears because of the impact by the plasma, but the CH plasma species appear in the SPNC process only.

5) At the treatment temperature of 713 K for 20 hours by such low temperature screen plasma technology, the SPN surface hardness was 850 Hv and 900 Hv in the case of plasma nitro-carburizing. In terms of depth of the hardened layer, the maximum hardness depth of SPNC was around twice deeper than SPN. That was proportional to the increased amount of nitrogen and carbon in the GDOES analysis.

REFERENCES

1 Sun, Y. and Bell, T. (1997) A Numerical Model of Plasma Nitriding of Low Alloy Steels. Materials Science and Engineering: A, 224, 33-47.

2 Tsujikawa, M., Yamauchib, N., Uedab, N., Soneb, T. and Hirosec, Y. (2004) Behavior of Carbon in Low Temperature Plasma Nitriding Layer of Austenitic Stainless Steel. Surface and Coatings Technology, 193, 309-313.

3 Lan, H.-Y. and Wen, D.-C. (2012) Improving the Erosion-Corrosion Properties of Precipitation Hardening Mold Steel by Plasma Nitriding. Materials Transactions, 53, 1443-1448.

4 Hubbard, P., Dowey, S.J., Partridge, J.G., Doyle, E.D. and McCulloch, D.G. (2010) Investigation of Nitrogen Mass Transfer within an Industrial Plasma Nitriding System II: Application of a Biased Screen. Surface and Coatings Technology, 204, 1151-1157.

5 Kim, S.G., Lee, J.H., Saito, N. and Takai, O. (2013) The Role of Activated Nitrogen Species on Double-Folded Screen Nitriding Process. Journal of Physics Conference Series, 417, 012-023.

6 Czerwiec, T., Renevier, N. and Michel, H. (2000) Low Temperature Plasma-Asisted Nitriding. Surface and Coatings Technology, 131, 267-277.

7 Michel, H., Czerwiec, T., Gantois, M., Ablitzer, D. and Ricard, A. (1995) Progress in the Analysis of the Mechanisms of Ion Nitriding. Surface and Coatings Technology, 72, 103-111.

8 Priest, J.M., Baldwin, M.J. and Fewell, M.P. (2001) The Action of Hydrogen in Low-Pressure r.f.-Plasma Nitriding. Surface and Coatings Technology, 145, 152-163.

9 Petitjean, L. and Ricard, A. (1984) Emission Spectroscopy Study of N2-H2 Glow Discharge for Metal Surface Nitriding. Journal of Physics D, 17, 919-929.

10 Gallo, S.C. and Dong, H. (2009) Study of Active Screen Plasma Processing Conditions for Carburising and Nitriding Austenitic Stainless Steel. Surface and Coatings Technology, 203, 3669-3675.

11 Leyland, A., Fancey, K.S., James, A.S. and Matthews, A. (1990) Enhanced Plasma Nitriding at Low Pressures: A Comparison between DC and RF Techniques. Surface and Coatings Technology, 41, 295-304.

12 Sharma, M.K., Saikia, B.K. and Bujarbarua, S. (2008) Optical Emission Spectroscopy of DC Pulsed Plasmas Used for Steel Nitriding. Surface and Coatings Technology, 203, 229-233.

13 Belmonte, T., Jaoul, C. and Borges, J.N. (2004) Modelling Nitrogen Atom Flux in Post-Discharge Nitriding Processes. Surface and Coatings Technology, 188-189, 201-206.

14 Musil, J., Vlcek, J. and Ruzicka, M. (2000) Recent Progress in Plasma Nitriding. Vacuum, 59, 940-951.

15 Pearse, R.W.B. and Gaydon, A.G. (1976) The Identification of Molecular Spectra. 4th Edition, Chapman and Hall, London/Wiley, New York.

CHAPTER 5

Electrochemical Investigation of Corrosion on AISI 316 Stainless Steel and AISI 1010 Carbon Steel: Study of the Behaviour of Imidazole and Benzimidazole as Corrosion Inhibitors

Roberta R. Moreira, Thiago F. Soares, Josimar Ribeiro*

Department of Chemistry, Federal University of Espírito Santo, Vitória, Brazil

ABSTRACT

An electrochemical investigation of the corrosion on AISI 316 austenitic stainless steel and AISI 1010 carbon steel in sodium chloride solution (3.0 wt.%) was performed in the absence and presence of imidazole and benzimidazole corrosion inhibitors. The results showed that at any inhibitor concentration (25 ppm to 1000 ppm), there was an increase in the polarisation resistance of both steels. The highest efficiency of corrosion inhibition was obtained using imidazole at a concentration of 50 ppm for both steels, with values of 96% for the AISI 316 stainless steel and 73% for the AISI 1010 carbon steel.

KEYWORDS

Stainless Steel, Carbon Steel, Polarisation, Anodic Dissolution, Inhibition

1. INTRODUCTION

Corrosion as a destructive phenomenon incurs a high cost to the economy and society. The costs have been estimated to be as high as 5% of GDP in many developed countries [1] [2] . All metallic material can undergo different types of degradation processes when exposed to different environmental conditions, either by natural factors or by industrial or other human activities [3]. Corrosion can affect critical areas or regions of a material and thereby modify its true functionality [4]. Additionally, the lifetime of metallic equipment tends to decrease further in more aggressive environments [1] .

In coastal regions, the corrosion of steel infrastructure can have considerable negative effects, and due to the high costs of maintenance and corrosion protection measures, there is interest in applying increasingly resistant steels [5] . According to Tsutsumi et al. [6] , sea water is an aggressive environment that affects almost all structural materials. Present in the composition of sea water are salts such as NaCl and $MgCl_2$, which are primary agents responsible for localised corrosion. The chloride ion is the most common in marine environments and the main cause of the breakdown of the passive film on steels [7] [8] . The presence or absence of this protective film on the surface of metals is responsible for controlling the behaviour of corrosion because of its action as a protective barrier against corrosion attacks [4] [9] .

Several metals suffer pitting corrosion when exposed to solutions with high chloride content [10]. Pitting corrosion is characterised by a process of localised metal dissolution that affects the integrity of the metal and, in some cases, leads to catastrophic failure. Although this form of corrosion occurs only in small areas of the metal surface, it remains a major concern due to its fast attack and penetration inside the metal [11] [12] .

The increased use of stainless steel has contributed substantially to the economic development of many countries. For decades, it has been used structurally in architectural elements such as beams and arches in bridges [13] .

In the case of stainless steel in contact with seawater, the corrosion rate and pitting potential depend on the levels of Cr and Ni present in the steel. The presence of Co, Mo and N has a significant influence on the corrosion resistance [14] . Austenitic stainless steels, whose main alloying elements are chromium and nickel, are considered to have high corrosion resistance in most environments compared with ferritic and martensitic steels [15] . The AISI 316 stainless steel, specifically, is perhaps one of the most used steels, after carbon steel, serving a structural material in industrial plants in marine environments, including desalination plants [14] . Carbon steel is also widely used in systems that operate in land and marine environments [16] and is found in various applications in our everyday life due to its relatively low cost [17] .

To preserve the integrity of metallic materials, some measures to inhibit or prevent corrosion in many aggressive media are cited in the literature [4] [18] . Organic inhibitors that exhibit one or more polar functions (such as N, O and S) and heterocyclic compounds with polar groups and π electrons have been quite effective in corrosion protection [19] . This inhibition efficiency is usually attributed to the specific interactions that occur between functional groups and heteroatoms with the metal surface due to their lone pair electrons [20] and the influence on the change of corrosion potential [20] [21] . Certain organic compounds, such as thiophene derivatives, were studied as corrosion inhibitors for AISI 316 stainless steel in acid media with chloride ions and resulted in an efficiency of approximately 97% [22] . According to studies conducted by Farahani and Goudarzi [23] , a benzothiazole derivative was also efficient in inhibiting the corrosion of AISI 316 stainless steel in an acid medium.

With respect to AISI 1010 carbon steel, a compound commercially known as Dodigen, which is a mixture of amines and amides, was studied as a corrosion inhibitor by Granero et al. [24] . The authors found that Dodigen had an efficiency ranging from 74% to 98%, depending on the concentration used. Other investigations using computer simulations of new imidazole derivatives showed that these compounds possess the ability to inhibit the corrosion process on AISI 1010 carbon steel, which showed 62.8% inhibition efficiency [25] .

Other organic inhibitors such as benzimidazole and its derivatives have been reported in the literature as effective corrosion inhibitors with structural features that allow adsorption to the metal surface [26] [27] . Benzimidazole, an imidazole derivative, is an organic compound that has aromaticity and nitrogen heteroatoms with lone pair electrons, which facilitate the adsorption to the metal [27] - [29] . Studies have shown that it provides a high corrosion inhibition ability to copper in acidic, neutral and alkaline media [26] and to carbon steel in acidic medium [30] .

Imidazole and its derivatives have been used as corrosion inhibitors for various metals and alloys [30] - [33] . The advantages of imidazole, including that it does not have a detrimental effect on the environment and its convenience and low cost, have made the use of these compounds very attractive [32] - [37] . Imidazole is a planar organic five-membered aromatic heterocycle that has two nitrogen atoms [31] . The protective mechanism of this molecule to the surface of the metallic material can occur via the adsorption of electron pairs isolated from the pyrrole nitrogen or parallel to the surface [36] of the imidazole ring.

This paper investigated the corrosion of AISI 316 austenitic stainless steel and AISI 1010 carbon steel in solutions containing 3.0 wt% chloride ions and the efficiency of corrosion inhibition of benzimidazole and imidazole inhibitors in different concentrations. The morphological and compositional analyses were performed by Laser-induced plasma emission spectroscopy (LIPS), X-ray diffraction (XRD), scanning electron microscopy (SEM) and energy dispersive X-ray (EDX), and the electrochemical study was performed by potentiodynamic polarisation.

2. EXPERIMENTAL SECTION

2.1. Sample Preparation

The samples used in this study were AISI 316 austenitic stainless steel, provided by Tecinox industry, and AISI 1010 carbon steel, provided by the Brazil Steel RDG; both companies are located in Espírito Santo, Brazil. The samples were prepared in two different ways. For morphological analysis, the samples were prepared in a square shape, with dimensions of 1.0 cm × 1.0 cm. For the electrochemical investigations, the samples were L-shaped, with dimensions of 3.0 cm × 1.0 cm. Each sample was prepared to leave an area of 1.0 cm^2 to test, which was isolated with the aid of an epoxy resin (Araldite®).

Before performing each analysis, the samples were subjected to cleaning and polishing of the surface using sandpaper of several different particle sizes in water, respectively, in the following order: 220, 320, 400, 600 and 1200 mesh. The polishing was finished with 0.3 µm alumina to obtain a uniform surface. At each change of sanding, the sample was rotated 90°. The specimens were then washed with distilled water, degreased with acetone and dried with a hot air blast. Before and after every analysis, the samples were stored in argon to prevent contamination by contact with the external environment.

2.2. Physico-Chemical Characterisation

Laser-induced plasma emission spectroscopy using a Foundry-Master Pro spectrometer of the Shimadzu Corporation equipped with an Echelle optics and Kodak KAF 1001 ICCD detector allowed us to estimate the chemical composition of the alloys.

The XRD analyses were carried out on a Bruker model D8 Discover diffractometer operating with Cu K_α radiation (λ = 1.5406 Å) generated at 40 kV and 40 mA; 2θ was scanned from 20° to 100° at a scan rate of 2° min^{-1}. The EVA V3.1 software was used to process the XRD data.

Metallographic analysis was performed to reveal the microstructure of steels. A chemical attack was performed on the sample of AISI 316 austenitic stainless steel with 1:3 HNO_3:HCl solution for 20 seconds to reveal the structure of the austenite grains and the presence of carbides. In the case of AISI 1010 carbon steel, 2% Nital (98% ethyl alcohol and 2% nitric acid) was used as the attack solution for 15 s to reveal the grain boundaries of ferrite. Microstructural characterisation was performed using a Nikon Eclipse 200 MA optical microscope at 1000×

magnification with polarised light. A Shimadzu SS550 scanning electron microscope coupled to an SEDX model analyser was used to investigate the corrosion morphology. The EDX results were obtained from the matrix interference, atomic number, absorbance, and fluorescence (ZAF correction) data. Between the electrochemical and morphological analyses, all of the samples were stored in argon atmosphere to avoid any outside contamination.

2.3. Electrochemical Characterisation

Electrochemical measurements were performed using a 100 mL capacity electrochemical cell with three electrodes. The reference electrode was $Ag/AgCl_{(KCl,\ 3mol/L)}$, the counter electrode was carbon with dimensions of 2.1 cm × 0.5 cm × 1.8 cm, and the working electrode was made of AISI 316 austenitic stainless steel or AISI 1010 carbon steel. For electrochemical tests, a potentiostat/galvanostat AUTOLAB PGSTAT model 302N and GPES software were used. The corrosion tests were performed in duplicate using reagents to prepare solutions and analytical grade deionised water with a conductivity of 23.07 $mS{\cdot}cm^{-1}$ at 25°C from a deioniser trademark of Union analysis. The temperature was monitored and remained at 24°C ± 1.0°C. The corrosive media used in the tests were composed of sodium chloride at 3.0 wt% using PA-Impex reagent ACS in the presence or absence of benzimidazole (Sigma-Aldrich, 98%) and/or imidazole (Sigma-Aldrich, 98%). Imidazole and benzimidazole were employed as corrosion inhibitors at different concentrations: 25, 50, 100, 500 and 1000 ppm. For Tafel polarisation measurements, the potential was varied by ±250 mV from the open circuit potential (E_{ocp}) at a scan rate of 0.5 $mV{\cdot}s^{-1}$. $E_{ocp}$was obtained by measuring electrochemical noise and was measured until the system came into balance for a period of 4000 seconds.

3. RESULTS AND DISCUSSION

3.1. Laser-Induced Plasma Emission Spectroscopy (LIPS)

Table 1 shows the results for the chemistry of the steels obtained via LIPS. The results obtained for the studied steels are within the composition range of AISI 316 [15] and AISI 1010 [38] as described in the literature.

3.2. X-Ray Diffraction

The XRD patterns of the analysed samples of AISI 316 stainless steel and AISI 1010 carbon steel are shown in Figure 1. The possible crystallographic phases present in each sample of steel and the orientation of these crystals can be observed. The XRD pattern of AISI 316 stainless steel shows diffraction peaks at 43.59°, 74.65° and 90.63°, relative to iron-nickel phase (FeNi) with reflection planes (111), (220) and (311), respectively, as the PDF-[01-077-7974]. A peak relative to chromium (Cr) phase with reflection plane (200) in accordance with the PDF-[01-088-2323] was observed at 50.78°. Both phases present have a face-centred cubic structure.

The XRD pattern of AISI 1010 carbon steel contains peaks that can be attributed to iron (Fe) present in a single phase composed of ferrite whose 2θ angles are 44.71°, 65.08° and 82.41° with reflection planes (110), (200) and (211), respectively, as the PDF-[06-0696]. The observed phase structure is body-centred cubic.

3.3. Physico-Chemical Investigation

The metallography images of the studied steels are presented in Figure 2 and show the grain boundaries of the phases present in the steel and the presence of carbides. The image obtained for the AISI 316 steel (Figure 2(a)) shows that the chemical attack revealed the grain boundaries of the austenite phase grains of precipitated carbides and the presence of twinned boundaries, which is caused by the displacement of atoms in the lattice due to stress or heat treatment.

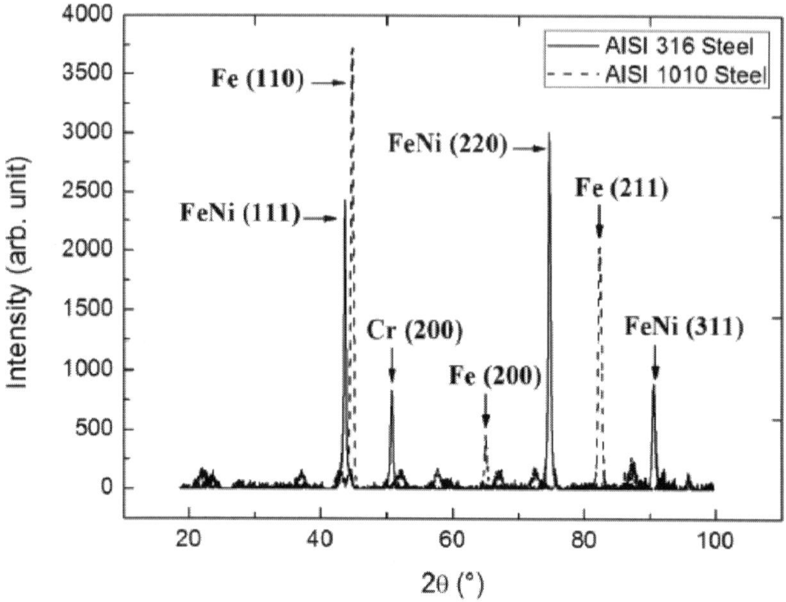

Figure 1. XRD pattern of AISI 316 stainless steel and AISI 1010 carbon steel.

Table 1. Chemical composition of AISI 316 austenitic stainless steel and AISI 1010 carbon steel.

AISI	Chemical Composition (wt%)							
	C	Mn	Si	Cr	Mo	P	S	Ni
316	0.0501	1.34	0.507	16.3	2.01	0.0278	0.0019	10.5
1010	0.0954	0.351	0.0083	0.0238	0.0023	0.0150	0.0084	0.0074

Similarly, the metallographic image of AISI 1010 steel (Figure 2(b)) revealed the grain boundaries present, which are composed of ferrite. Higher than the amounts of AISI 316 were also revealed that grain carbide steel, including regions of the grain boundary, which can be observed in Figure 2(b). Such regions, including the steel grain boundaries, deformation and non-metallic inclusions, have been found in the literature to be preferred sites for the occurrence of corrosion, particularly pitting corrosion [16] [39] .

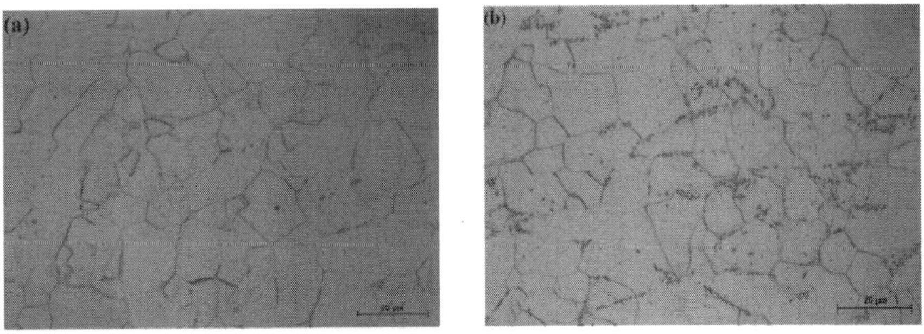

Figure 2. Metallographic images at 1000× magnification of the two grades of steel before the attack of chloride ions in solution. (a) AISI 316 stainless steel and (b) AISI 1010 carbon steel.

The scanning electron microscopy and EDX spectra of the steel before the etching process are shown in Figure 3. Nonmetallic inclusions are present in both of the steel samples studied. These inclusions are more prone to the start of an anodic attack in a corrosive environment, and these areas become the sites of nucleation of pitting [16] [39] . The difference in composition between regions of inclusions and the matrix of the steel is responsible for generating the potential difference in these two areas, which initiates the corrosion process.

The EDX analysis of AISI 316 (Figure 3(b)) stainless steel shows the elemental composition of the sample found in the possible inclusion of steel and the elemental composition of the matrix of the steel. From this analysis, it is observed that this inclusion in the AISI 316 steel is composed mostly of carbon and oxygen. The EDX analysis also shows peaks related to silicon, manganese, sulphur and chromium in lower content than the matrix, showing that it can accommodate Si and Cr oxides, as well as the presence of carbides and manganese sulphide. The presence of grains of carbides precipitated in the steel is in accordance with the analytical results of the inclusions performed using the metallography technique, as mentioned above.

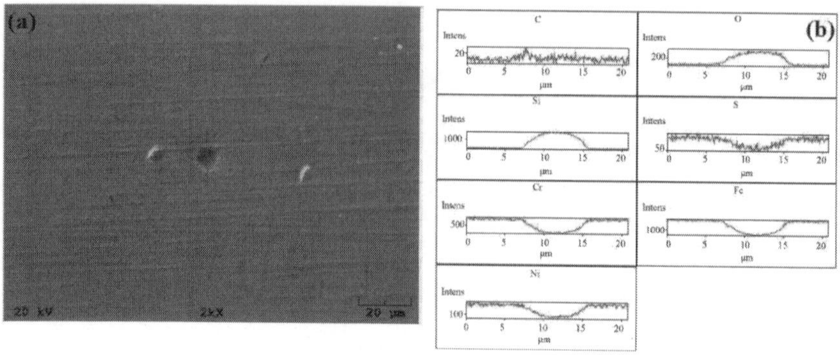

Figure 3. SEM images (a) and EDX analyses (b) of sample of AISI 316 steel at 2000× magnification before the corrosion process.

Elements such as oxygen, sulphur, nitrogen, phosphorus, chlorine, sodium and potassium may exist in steel primarily as a result of the steel-making process and are responsible for generating negative effects on the properties of steels, causing brittleness and lower corrosion resistance. To reduce the amount of oxygen and sulphur in Al-Si alloy, for example, elements are added in the metal degreasing process because of the strong reactivity of these elements to oxygen forming their oxides [40] .

Figure 4 shows EDX spectra and micrographs of the steels after the corrosion process in a solution containing 3.0 wt% chloride ions. The SEM analysis of the surface of the steels after suffering corrosive attack by chloride ions allowed the identification of the type of corrosion that occurred on the surface of each of the steels. After analysing the image in Figure 4(a), which shows the surface of the AISI 316 steel after attack by chloride ions, the type of corrosion was identified as the pitting type, which is characterised by a larger corrosion depth than diameter and occurs in small, localised regions of the metal surface. However, the image that shows the AISI 1010 steel surface after attack by chloride ions (Figure 4(b)) reveals that there has been widespread and uniform corrosion on the steel surface because the corrosion is present across the extent of the surface and because the corrosion product exhibits porosity.

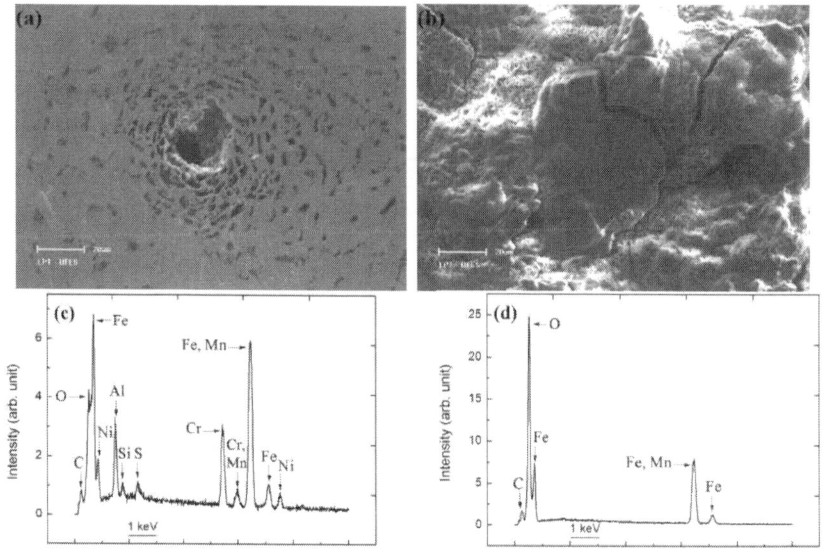

Figure 4. SEM images and EDX analyses of steel samples at 600× magnification after the corrosion process in the presence of sodium chloride at 3.0 wt.% solution. (a) (c) AISI 316 stainless steel; (b) (d) AISI 1010 carbon steel.

The corrosive attack of chloride ions to the AISI 316 steel was expected because pitting corrosion occurs in passive metals and steel in regions where there are flaws in the protective film, as observed in studies by Burstein and Pistorius [10] , Kondo [11] and Turnbull et al. [12] , and in deformed regions and non-metallic inclusions [16] [39] . This explanation is supported by SEM surface analysis and by EDX analysis of the pitting, which shows the presence of elements related to non-metallic inclusions (Figure 4(c)) and that this steel is susceptible to such corrosion characteristics. Conversely, because AISI 1010 steel is not passivated, the entire surface is exposed to the corrosive medium and is thus susceptible to widespread corrosion, instead of localised pitting corrosion, as observed in the surface analysis. Additionally, the EDX results (Figure 4(d)) contain no evidence of inclusions containing Al, Si or S, thereby reducing the likelihood of localised pitting corrosion.

3.4. Electrochemical Investigations

The electrochemical behaviour of AISI 316 stainless steels and AISI 1010 carbon steel in the presence of chloride ions was investigated using Tafel plots, as shown in Figure 5. From Tafel analysis of the curves, it was possible to obtain parameters related to the corrosion of steels, such as corrosion potential (E_{corr}) and polarisation resistance (R_p), which are important for the evaluation of the corrosion data. Anode active regions in which the current density increases with increasing potential are observed inFigure 5 and are consistent with the behaviour of steels. For steel AISI 316, this region begins at 0.15 V vs. $Ag/AgCl_{(KCl,\ 3\ mol/L)}$, and for AISI 1010 steel, it begins at −0.45 V vs. $Ag/AgCl_{(KCl,\ 3mol/L)}$.

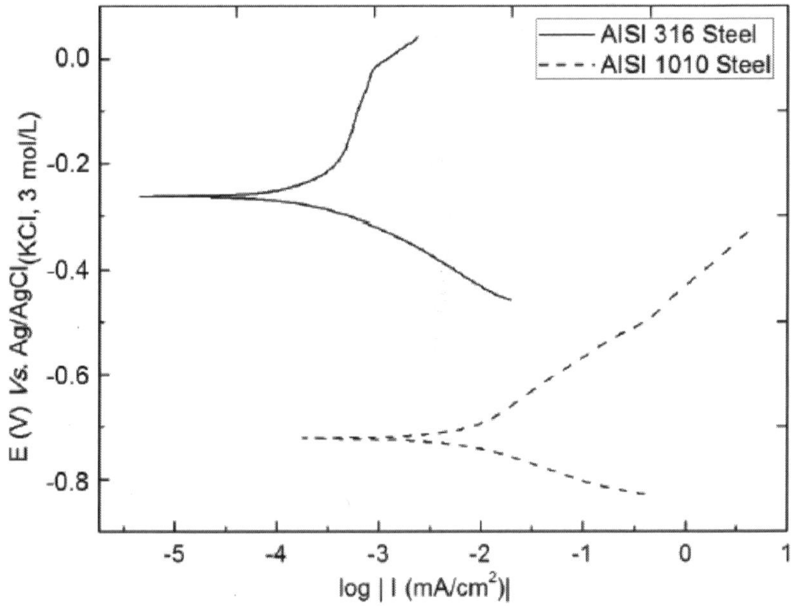

Figure 5. Tafel plots of the AISI 316 stainless steel and AISI 1010 carbon steel in the presence of 3.0 wt% solution of chloride ions. Temperature = 24°C ± 1.0°C; scan rate of 0.5 $mV{\cdot}s^{-1}$.

Furthermore, the current density increases more sharply with the increase of the potential for AISI 1010 steel, indicating that this steel has less resistance to corrosion steel AISI 316. This finding is in agreement with the results from the literature showing that the composition of steel influences the corrosion behaviour [14] [15] . This active behaviour is caused by the dissolution of the steel in the presence of chloride ion. The steel reacts to form the ferrous chloride, which is hydrolysed to form ferrous hydroxide and hydrochloric acid. This process decreases the pH of the medium and accelerates the corrosion mechanism [41] . The curves also show a region of potential where capacitive capacitive current exists, which is caused by the charging of the electric double layer. In this region, there is no increase in current density with increasing potential and virtually zero total current, which indicates that there is no breaking of the passive layer present on the steel surface [10] . For AISI 316, that potential range is approximately 1350 mV and is greater than that of AISI 1010 steel, which has an approximately 750 mV range. Thus, the AISI 316 steel has a more stable passive layer (data not shown).

The Tafel plots also show the corrosion resistance of AISI 316 steel compared with AISI 1010 steel, with −262 mV from E_{corr} vs. $Ag/AgCl_{(KCl,\ 3\ mol/L)}$ and −722 mV vs. $Ag/AgCl_{(KCl,\ 3\ mol/L)}$, respectively, in accordance with the analysis of potentiodynamic curves. This behaviour is due to the composition of each steel and the corrosion resistance imparted to the AISI 316 steel because of the presence of elements such as chromium and nickel, which are responsible for reducing the susceptibility of steel corrosion due to the possibility of film formation on the steel surface [14] [15] .

3.5. Electrochemical Study of Steels with Benzimidazole Inhibitor

Tafel curves for the AISI 316 and AISI 1010 steels in 3.0 wt% chloride ion solution + benzimidazole inhibitor at different concentrations (from 25 ppm to 1000 ppm) are shown in Figure 6. By extrapolating the Tafel plots, values of the electrochemical parameters associated with corrosion processes, such as corrosion potential (E_{corr}), corrosion current density (i_{corr}) and polarisation resistance (R_p), were obtained. The results of the corrosion inhibition efficiency of the benzimidazole were calculated from the values of the polarisation resistance (R_p) using Equation (1), as shown in the literature [42] .

$$n(\%) = \frac{Rp(\text{inhibitor}) - Rp}{Rp(\text{inhibitor})} \times 100 \quad (1)$$

R_p (inhibitor) and R_p are the polarisation resistance values in the presence or absence of inhibitor, respectively. Table 2 lists the obtained parameters and the corrosion inhibition efficiency (n) of benzimidazole.

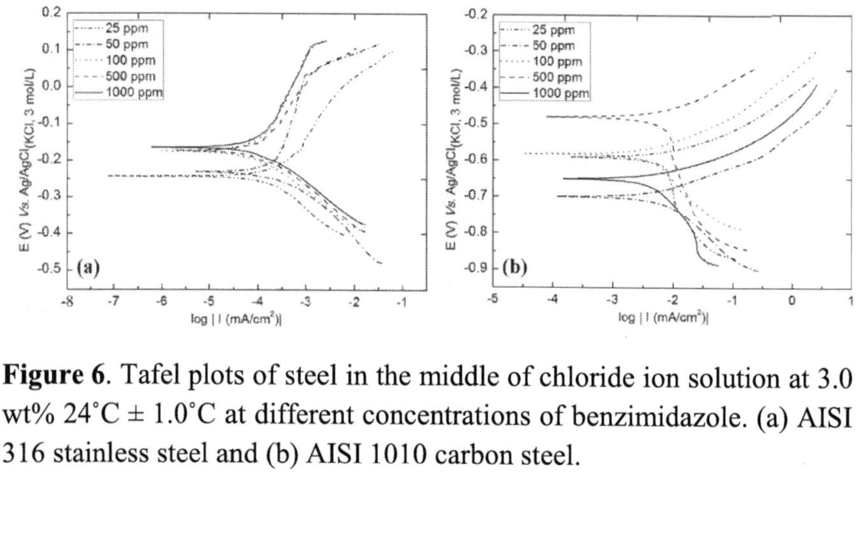

Figure 6. Tafel plots of steel in the middle of chloride ion solution at 3.0 wt% 24°C ± 1.0°C at different concentrations of benzimidazole. (a) AISI 316 stainless steel and (b) AISI 1010 carbon steel.

Table 2 shows that there was an increase in R_p for all concentrations of benzimidazole for both steels, which means that the inhibitor is inhibiting the corrosion reaction. Moreover, the corrosion potential (E_{corr}) shifted to more anodic potentials. For both steels, the best inhibition efficiency was observed at an inhibitor concentration of 100 ppm, with an efficiency of 71.4% and R_p = 8.93 kΩ·cm^2 for AISI 316 and an efficiency of 51.2% and R_p = 178.60 Ω·cm^2 for the AISI 1010 steel.

Table 2. Results obtained from Tafel plots for AISI 316 stainless steel and AISI 1010 carbon steel immersed in 3.0 wt% chloride ion solution at 24°C ± 1.0°C in the presence of different concentrations of benzimidazole (from 0 ppm to 1000 ppm).

Benzimidazole	AISI 316			AISI 1010		
	E_{corr} (mV)	Rp (kΩ·cm²)	η (%)	E_{corr} (mV)	Rp (Ω·cm²)	η (%)
0 ppm	−262	2.55	–	−722	87.14	–
25 ppm	−243	4.28	40.42	−591	89.29	2.41
50 ppm	−231	3.78	32.54	−700	142.90	39.02
100 ppm	−174	8.93	71.44	−582	178.60	51.21
500 ppm	−173	7.15	64.34	−480	110.50	21.14
1000 ppm	−165	5.53	53.89	−650	94.58	7.87

Khaled [27] used benzimidazole as a corrosion inhibitor for iron in acidic medium and showed that at a concentration of 0.05 mol·L^{-1} benzimidazole, which is greater than the concentration used in this paper, the maximum efficiency was 51%. This result shows that the benzimidazole effectively acted as an inhibitor for both the steel and the AISI 316 and AISI 1010 steels in corrosive environment studied. Tang et al. [42] used benzimidazole derivatives, such as 2-aminomethyl benzimidazole, as corrosion inhibitors for mild carbon steel in acid medium and showed a maximum corrosion inhibition efficiency of 74% at the highest concentration studied (0.002 mol·L^{-1}). Because the molecular structure of this inhibitor provides more adsorption sites compared with benzimidazole and is a molecule of greater steric body that can better cover the surface of the steel, it should be noted that benzimidazole showed a maximum inhibition efficiency of 71.4% for AISI 316 steel, which is very close to the maximum obtained by Tang et al. [42]. As observed in this study, benzimidazole showed a better inhibition for AISI 316 steel, and this behaviour was observed in almost all of the inhibitor concentrations studied.

According to the literature, benzimidazole derivatives protect the steel surface by adsorption to form an insoluble complex with Fe (II) and the segments of these derivatives that are longer than benzimidazole increase the molecule inhibition efficiency [42]. The adsorption mechanism is due to the lone pair electrons and the π electrons of the aromatic ring that are present in the inhibitor molecule and occupy the empty d orbitals of Fe atoms, forming a protective film on the steel surface [43] [44] . It is also possible that inhibitor packaging products of corrosion may hinder the diffusion of aggressive species to the steel surface [42]. Such phenomena may have been responsible for the results of corrosion inhibition by benzimidazole. However, the lower inhibition efficiency obtained for AISI 1010 steel compared with AISI 316 steel suggests that the benzimidazole may have failed to adsorb to the surface of this steel, which indicates a less effective and/or weaker adsorption onto the surface. Another suggestion may be the resulting geometry of the packing of inhibitor molecules, which leaves portions of the surface exposed to the corrosive environment. Additionally, the smaller steric body of this molecule may not have protected the entire surface. For example, in studies conducted by Khaled and Amin [43] using the thiazole derivative 2-Amino-4-(p-tolyl) thiazole, which has a larger steric body than benzimidazole, the maximum inhibition efficiency was 95.2% for carbon steel.

3.6. Electrochemical Study of Steels with Imidazole Inhibitor

Figure 7 shows the Tafel plots of AISI 316 and AISI 1010 steels obtained from the solution of 3.0 wt% chloride ions in the presence of imidazole inhibitor at different concentrations (from 25 ppm to 1000 ppm). The Tafel plots show that there is a decrease in the current density in the presence of imidazole inhibitor. Moreover, there was a change of corrosion potential (E_{corr}) to more anodic values.

The obtained parameters and the corrosion inhibition efficiency (n) of imidazole are shown in Table 3. For both steels, the inhibitor concentration that showed the best inhibition efficiency was 50 ppm, with an efficiency of ~96%, and R_p = 63.18 kΩ·cm^2 for the AISI 316 stainless steel and an efficiency of ~73% and R_p = 325.10 Ω·cm^2 for the AISI 1010 carbon steel.

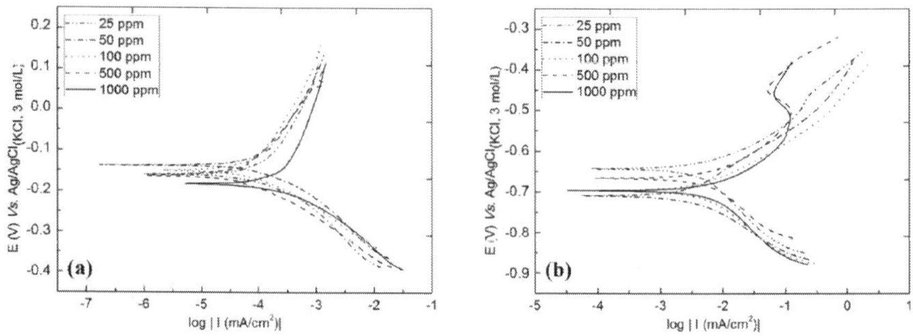

Figure 7. Tafel plots of steel in 3.0 wt% chloride ion solution at 24°C ± 1.0°C with different concentrations of imidazole. (a) AISI 316 stainless steel and (b) AISI 1010 carbon steel.

Table 3. Results obtained from Tafel plots for AISI 316 and AISI 1010 steels immersed in 3.0 wt% chloride ion solution at 24°C ± 1.0°C in the presence of different concentrations of imidazole (from 0 ppm to 1000 ppm).

Imidazole	AISI 316			AISI 1010		
	E_{corr} (mV)	Rp (kΩ·cm²)	η (%)	E_{corr} (mV)	Rp (Ω·cm²)	η (%)
0 ppm	−262	2.55	–	−722	87.14	–
25 ppm	−160	15.96	84.02	−643	187.20	53.45
50 ppm	−138	63.18	95.96	−709	325.10	73.20
100 ppm	−150	13.60	81.25	−696	176.70	50.68
500 ppm	−164	18.92	86.52	−666	101.00	13.72
1000 ppm	−183	3.02	15.56	−696	113.40	23.16

Zhang et al. [31] , using imidazole as a corrosion inhibitor for iron in acidic medium, achieved a maximum efficiency of 79.6% at a concentration of 0.1 mol·L^{-1} inhibitor, which was higher than the concentration used in this study here. Another study by He et al. [45] that used imidazole to inhibit corrosion in aluminium found a maximum efficiency of 79.4% at an imidazole concentration of 0.014 mol·L^{-1}. These results indicate that the imidazole proved effective in inhibiting corrosion for the two steels studied in this work and that higher inhibition efficiencies were obtained for AISI 316 stainless steel compared with

AISI 1010 carbon steel at almost all of the inhibitor concentrations studied.

The results showed that imidazole showed better corrosion inhibition efficiency than did benzimidazole. The imidazole molecule has a planar aromatic ring containing two N heteroatoms. According to Mendes et al. [36] , this molecule can be adsorbed by the imidazole ring, parallel to the surface of the metallic material, which may be one reason for the better efficiency of inhibition. Thus, it is suggested the imidazole molecules covered more of the metal surface due to a better compression between them, thus resulting in higher corrosion inhibition efficiency.

4. CONCLUSIONS

The analysis performed using XRD showed that the different phases are present in the steels and affect corrosion resistance. The SEM and EDX techniques make it possible to identify and determine the composition of the regions that are more susceptible to corrosion, such as inclusions in the steels. These techniques also allow for the further characterisation of the type of corrosion suffered by each steel.

Electrochemical studies have shown that at any inhibitor concentration (25 ppm to 1000 ppm), there was an increase of the polarisation resistance of both steels, indicating that there was an increase in corrosion resistance.

The best corrosion inhibition result for AISI 316 steel and AISI 1010 steel using benzimidazole was obtained at an inhibitor concentration of 100 ppm, and the inhibition efficiency values were ~71% and ~51%, respectively. When imidazole was used as the inhibitor, the best corrosion inhibition was obtained for both steels using imidazole at an inhibitor concentration of 50 ppm, with inhibition efficiency values of ~96% and ~73%, respectively.

ACKNOWLEDGEMENTS

The authors thank FAPES Nº 61612626/2013, CAPES, CNPq, UFES, PETROBRAS, IFES, and LPT-UFES TRICORRMAT-UFES.

REFERENCES

1 Groysman, A. and Brodsky, N. (2006) Corrosion and Quality. Accreditation and Quality Assurance, 10, 537-542.

2 Almeida, M. (2005) Minimisation of Steel Atmospheric Corrosion: Updated Structure of Intervention. Progress in Organic Coatings, 54, 81-90.

3 Hao, L., Zhang, S., Dong, J. and Ke, W. (2011) Atmospheric Corrosion Resistance of MnCuP Weathering Steel in Simulated Environments. Corrosion Science, 53, 4187-4192.

4 Shifler, D.A. (2005) Understanding Material Interactions in Marine Environments to Promote Extended Structural Life. Corrosion Science, 47, 2335-2352.

5 Kage, I., Matsui, K. and Kawabata, F. (2005) Minimum Maintenance Steel Plates and Their Application Technologies for Bridge-Life Cycle Cost Reduction Technologies with Environmental Safeguards for Preserving Social Infrastructure Assets. JFE Technical Report, 5, 31-37.

6 Tsutsumi, Y., Nishikata, A. and Tsuru, T. (2007) Pitting Corrosion Mechanism of Type 304 Stainless Steel under a Droplet of Chloride Solutions. Corrosion Science, 49, 1394-1407.

7 Laycock, N.J. and Newman, R.C. (1997) Localised Dissolution Kinetics, Salt Films and Pitting Potentials. Corrosion Science, 39, 1771-1790.

8 Newman, R.C. (2001) Understanding the Corrosion of Stainless Steel. Corrosion, 57, 1030-1041.

9 Martini, E.M.A. and Muller, I.L. (2000) Characterization of the Film formed on Iron in Borate Solution by Electrochemical Impedance Spectroscopy. Corrosion Science, 42, 443-454.

10 Pistorius, P.C. and Burstein G.T. (1992) Metastable Pitting Corrosion of Stainless Steel and the Transition to Stability. Philosophical Transactions of the Royal Society A, 341, 531-559.

11 Kondo, Y. (1989) Prediction of Fatigue Crack Initiation Life Based on Pit Growth. Corrosion, 45, 7-11.

12 Turnbull, A., Mccartney, L.N. and Zhou, S. (2006) A Model to Predict the Evolution of Pitting Corrosion and the Pit-to-Crack Transition Incorporating Statistically Distributed Input Parameters. Corrosion Science, 48, 2084-2105.

13 Baddoo, N.R. (2008) Stainless Steel in Construction: A Review of Research, Applications, Challenges and Opportunities. Journal of Constructional Steel Research, 64, 1199-1206.

14 Malik, A.U., Siddiqi, N., Ahmad, S. and Andijani, I.N. (1995) The Effect of Dominant Alloy Additions on the Corrosion Behaviour of Some Conventional and High Alloy Stainless Steels in Seawater. Corrosion Science, 37, 1521-1535.

15 Lippold, J.C. and Kotecki, D.J. (2005) Welding Metallurgy of Stainless Steel. Wiley-Intercience, New Jersey.

16 Avci, R., Davis, B.H., Wolfenden, M.L., Beech, I.B., Lucas, K. and Paul, D. (2013) Mechanism of MnS-Mediated Pit Initiation and Propagation in Carbon Steel in an Anaerobic Sulfidogenic Media. Corrosion Science, 76, 267-274.

17 Rangaraju, R.R., Raja, K.S., Panday, A. and Misra, M. (2010) Low-Cost Photoelectrocatalyst Based on a Nanoporous Oxide Layer of Low-Carbon Steel. Journal of Physics D: Applied Physics, 43, Article ID: 445301.

18 Ormellese, M., Lazzari, L., Goidanich, S., Fumagalli, G. and Brenna, A. (2009) A Study of Organic Substances as Inhibitors for Chloride-Induced Corrosion in Concrete. Corrosion Science, 51, 2959-2968.

19 Fekry, A.M. and Ameer, M.A. (2010) Corrosion Inhibition of Mild Steel in Acidic Media Using Newly Synthesized Heterocyclic Organic Molecules. International Journal of Hydrogen Energy, 35, 7641-7651.

20 Lunarska, E. and Chernyayeva, O. (2006) Effect of Corrosion Inhibitors on Hydrogen Uptake by Al from NaOH Solution. International Journal of Hydrogen Energy, 31, 285-293.

21 Zucchi, F., Trabanelli, G., Frignani, A. and Zucchini, M. (1978) The Inhibition of Stress Corrosion Cracking of Stainless Steels in Chloride Solutions. Corrosion Science, 18, 87-95.

22 Galal, A., Atta, N.F. and Al-Hassan, M.H.S. (2005) Effect of Some Thiophene Derivatives on the Electrochemical Behaviour of AISI 316 Austenitic Stainless Steel in Acidic Solutions Containing Chloride Ions I. Molecular Structure and Inhibition Efficiency Relationship. Materials Chemistry and Physics, 89, 38-48.

23 Goudarzi, N. and Farahani, H. (2014) Investigation on 2-Mercaptobenzothiazole Behaviour as Corrosion Inhibitor for 316-Stainless Steel in Acidic Media. Anti-Corrosion Methods and Materials, 61, 20-26.

24 Granero, M.F.L., Matai, P.H.L.S., Aoki, I.V. and Guedes, I.C. (2009) Dodigen 213-N as Corrosion Inhibitor for ASTM 1010 Mild Steel in 10% HCl. Journal of Applied Electrochemistry, 39, 1199-1205.

25 Duda, Y., Govea-Rueda, R., Galicia, M., Beltrán, H.I. and Zamudio-Rivera, L.S. (2005) Corrosion Inhibitors: Design, Performance and Computer Simulations. The Journal of Physical Chemistry B, 109, 22674-22684.

26 Kuznetsov, Y.I. and Kazansky, L.P. (2008) Physicochemical Aspects of Metal Protection by Azoles as Corrosion Inhibitors. Russian Chemical Reviews, 77, 219-232.

27 Khaled, K.F. (2003) The Inhibition of Benzimidazole Derivatives on Corrosion of Iron in 1M HCl Solutions. Electrochimica Acta, 48, 2493-2503.

28 Wang, L. (2001) Evaluation of 2-Mercaptobenzimidazole as Corrosion Inhibitor for Mild Steel in Phosphoric Acid. Corrosion Science, 43, 2281-2289.

29 Popova, A., Christov, M. and Deligeorgiev, T. (2003) Influence of the Molecular Structure on the Inhibitor Properties of Benzimidazole Derivatives on Mild Steel Corrosion in 1 M Hydrochloric Acid. Corrosion, 59, 756-764.

30 Aljourani, J., Raeissi, K. and Golozar, M.A. (2009) Benzimidazole and Its Derivatives as Corrosion Inhibitors for Mild Steel in 1M HCl Solution. Corrosion Science, 51, 1836-1843.

31 Zhang, Z., Chen, S., Li, Y., Li, S. and Wanga, L. (2009) A Study of the Inhibition of Iron Corrosion by Imidazole and Its Derivatives Self-Assembled Films. Corrosion Science, 51, 291-300.

32 Curkovic, H.O., Stupnišek-lisac, E. and Takenouti, H. (2010) The Influence of pH Value on the Efficiency of Imidazole Based Corrosion Inhibitors of Copper. Corrosion Science, 52, 398-405.

33 Zhang, D.Q., Gao, L.X. and Zhou, G.D. (2004) Inhibition of Copper Corrosion in Aerated Hydrochloric Acid Solution by Heterocyclic Compounds Containing a Mercapto Group. Corrosion Science, 46, 3031-3040.

34 Popova, A., Sokolova, E., Raicheva, S. and Christov, M. (2003) AC and DC Study of the Temperature Effect on Mild Steel Corrosion in Acid Media in the Presence of Benzimidazole Derivatives. Corrosion Science, 45, 33-58.

35 Kokalj, A. (2013) Formation and Structure of Inhibitive Molecular Film of Imidazole on Iron Surface. Corrosion Science, 68, 195-203.

36 Mendes, J.O., Silva, E.C. and Rocha, A.B. (2012) On the Nature of Inhibition Performance of Imidazole on Iron Surface. Corrosion Science, 57, 254-259.

37 Mousavi, M., Mohammadalizadeh, M. and Khosravan, A. (2011) Theoretical Investigation of Corrosion Inhibition Effect of Imidazole and Its Derivatives on Mild Steel Using Cluster Model. Corrosion Science, 53, 3086-3091.

38 ASM International Handbook Committee (1990) Volume 1: Properties and Selection: Irons, Steels and High-Perfor- mance Alloys. ASM International, Ohio, 150.

39 Li, Y., Hu, R.G., Wang, J.R., Huang, Y.X. and Lin, C.-J. (2009) Corrosion Initiation of Stainless Steel in HCl Solution Studied Using Electrochemical Noise and in Situ Atomic Force Microscope. Electrochimica Acta, 54, 7134-7140.

40 Gammer, K., Rosner, M., Poeckl, G. and Hutter, H. (2003) AES and SIMS Analysis of Non-Metallic Inclusions in a Low-Carbon Chromium-Steel. Analytical and Bioanalytical Chemistry, 376, 255-259.

41 Yang, Y.Z., Jiang, Y.M. and Li, J. (2013) In Situ Investigation of Crevice Corrosion on UNS S32101 Duplex Stainless Steel in Sodium Chloride Solution. Corrosion Science, 76, 163-169.

42 Tang, Y., Zhang, F., Hu, S., Cao, Z., Wu, Z. and Jing, W. (2013) Novel Benzimidazole Derivatives as Corrosion Inhibitors of Mild Steel in the Acidic Media. Part I: Gravimetric, Electrochemical, SEM and XPS studies. Corrosion Science, 74, 271-282.

43 Khaled, K.F. and Amin, M.A. (2009) Corrosion Monitoring of Mild Steel in Sulphuric Acid Solutions in Presence of Some Triazole Derivatives—Molecular Dynamics, Chemical and Electrochemical Studies. Corrosion Science, 51, 1964-1975.

44 Negma, N.A., Kandile, N.G., Badr, E.A. and Mohammed, M.A. (2012) Gravimetric and Electrochemical Evaluation of Environmentally Friendly Nonionic Corrosion Inhibitors for Carbon Steel in 1 M HCl. Corrosion Science, 65, 94-103.

45 He, X., Jiang, Y., Li, C., Wang, W., Hou, B. and Wu, L. (2014) Inhibition Properties and Adsorption Behaviour of Imidazole and 2-Phenyl-2-Imidazoline on AA5052 in 1.0 M HCl Solution. Corrosion Science, 83, 124-136.

CHAPTER 6

Effects of Gas Nitriding on the Mechanical and Corrosion Properties of SACM 645 Steel

Shu-Hung Yeh[1], Liu-Ho Chiu[1], Heng Chang[2]

[1]40, Chung-Shan North Road, 3rd Sec., Chung-Shan District, Taipei City, Chinese Taipei

[2]55, Hwa Kang Road, Yang Ming Shan, Taipei, Chinese Taipei

ABSTRACT

The effects of the nitrided case produced by gas nitriding processes on the mechanical and corrosion resistance properties of the JIS SACM 645 steel were studied in this paper. JIS SACM 645 steel specimens with different substrate hardness were gas nitrided at 530˚C for various nitriding durations. Nitrided specimens were characterized by means of optical and scanning electron microscopy, X-ray diffraction, glow discharge optical spectrometry, microhardness profiling, wear test, torsion mode fatigue test as well as electrochemical corrosion test in an aerated 3.5% NaCl solution. The surface hardness values of the nitrided specimens with Fe_3N and Fe_4N phases precipitated in the case layer were observed higher than 1000 $HV_{0.1}$. Mass loss measurement of the wear test showed increases of wear resistance of the nitrided specimens, and the mass losses of the specimens were strongly influenced by nitriding durations. Electrochemical measurements showed that corrosion current density of the specimens was significantly decreased after nitriding and the corrosion potential was shifted to the noble direction as the increase of the nitriding durations. The fatigue limit of the specimen nitrided for 96 h rose 44% to 600 MPa in exceeding the untreated specimen in this study.

KEYWORDS

Nitriding, Steel, Wear, Fatigue

1. INTRODUCTION

The positive effects of surface treatments like nitriding on the fatigue and wear properties of structural steel have been known for several decades. Tools and components used in the severe environment must have excellent fatigue resistance and wear resistance. Consequently, the structural steel of those components should be always surface modified to provide these needed properties. The surfaces of these components are required to attain high hardness values to resist wear, and the inner cores still need to have high toughness strength [1,2]. Therefore, various methods, such as nitriding, carburizing, ceramic coating and thermal chemical treatments, have been devised to improve the surface mechanical properties of the components [3-5]. Among those surface treatment processes, the gas nitriding methods have been used extensively to enhance the mechanical properties of steel components [5-13]. During the gas nitriding treatment, nitrogen, which is provided by the catalytic dissociation of ammonia, diffuses through the steel surface to form a concentration gradient as well as nitride precipitates in the nitriding case. The solid solution of nitrogen in the metal substrate and the nitride precipitates are responsible for the increase of hardness and compressive macrostresses in the diffusion layer [7].

The fatigue characteristics of nitrided SACM 645 steel were investigated by Bell et al. [5]. According to the nitriding condition and notch shape, the improvement of the fatigue characteristics of the notched specimens was more effective than unnotched specimens. Peng [11] investigated ammonia-gas nitriding on AISI 304 austenitic stainless steel at temperatures higher than 800°C. The result from Peng showed that a metas phase named 'S-phase' or 'γ_N-phase' (expanded austenite) was formed at such high temperatures due to a sufficiently high nitriding potential of ammonia gas. Ashrafizadeh [12] further concluded that the compound layer had no significant influence on fatigue limits, but instead it was the diffusion layer which had the dominating effect on the fatigue behavior of nitrided specimens. Pellizzari [13] also reported that the hard porous compound layer was brittle and favored crack initiation to decrease the thermal fatigue resistance. As the reasons for the improvement in fatigue limit have not been known, there are also lack of studies on the quantitative relations between process parameters of the surface treatment and the fatigue limits. The aim of present study is to investigate the effects of different gas nitriding parameters on the mechanical and corrosion properties of the SACM 645 steel, as well as the effect of the substrate hardness on the wear, fatigue and corrosion properties of nitrided SACM 645 steel specimens.

2. EXPERIMENTAL PROCEDURES

The chemical composition (wt%) of the specimen analyzed by a glow-discharge optical emission spectroscope (LECO, GDOES-750) is listed in **Table 1**. The specimens were quenched and tempered to the substrate hardness of 30, 35 and 40 HRC before nitriding. The surfaces of specimens were fine-ground using abrasive papers to a 1200-grit finish and cleaned with acetone solution in an ultrasonic cleaner. Treatments were carried out in a gas nitriding furnace and the process parameters used in the nitriding of steel was set at 530°C for 12 h, 24 h, 48 h, 96 h. The chamber atmosphere was filled with NH_3 gas at a flow rate of 0.6 ml/min. To classify the specimens undergone various treatment conditions, specimen abbreviations are coded in **Table 2**.

The cross-sectional microstructure of the specimens was revealed using a chemical solution of 95 ml alcohol +5 ml HNO_3, and was observed on an optical microscope (Nikon OPTIPHOT-100). The fracture surface of the nitrided specimen through fatigue test, conducted on a torsion mode machine with 3600 rpm at room temperature, was examined on a JEOL JSM-5600 scanning electron microscope (SEM). The phases precipitated in the case layer of the nitrided specimen were characterized on an X-ray diffractometer with a Cu-Kα radiation source over the 2θ range from 20°

to 120°. Microhardness was measured by using a Vickers microhardness tester with the applied load of 100 g. The wear resistance of the nitrided specimens was investigated on a non-lubricated block-on-roller tester. The counter-roller on the block-on-roller tester was made of JIS SKS3 (AISI O1) tool steel with hardness of 62HRC. A normal load of 95.9 N was applied on the roller and the counter-roller was rotated at 180 rpm. Fatigue tests were carried out in torsion mode at a stress reversal frequency of 3600 rpm. The fatigue limit was determined by the so-called staircase method and was defined at 10^7 cycles. The corrosion behavior was evaluated by electrochemical analysis. Potentiodynamic polarization measurements were carried out using a conventional three-electrode cell housing a working electrode, a platinum counter electrode, and a saturated silver chloride (SSC) reference electrode in a 3.5% NaCl solution at 25°C. The potentiodynamic scan was recorded starting from corrosion potential of -0.6 V to 1.0 V in a potential scan rate of 1 mV/s.

Table 1. Chemical composition of the SACM 645 steel (wt%).

Fe	C	Si	Mn	Cr	Mo	Al
96.05	0.482	0.216	0.529	1.656	0.164	1.013

Table 2. Codes of specimens and treatments.

code	Treatment
X-TMS	Quenched and tempered to X HRC
XC-GNY	Substrate hardness of X HRC + Gas nitrided at 530°C for Y hours

3. RESULTS AND DISCUSSION

3.1. Microstructures

Figure 1 shows the cross sectional micrographs of nitrided specimens produced by the gas nitriding treatment at 530°C of different time durations. The specimens were mechanically fine-ground and polished, followed by etching in a 5% Nital solution. A slightly etched outer white compound layer of approximately 15 mm was observed and were soundly bonded with the nitrogen diffusion layer underneath. Moreover, compound layer without voids and porosities was obtained for all nitrided specimens with the substrate hardness of 30 HRC, even as the nitriding duration prolonging to 96 h. In comparison, the compound layer of specimen with substrate hardness of 40 HRC exhibited a non-homogeneous structure as shown in **Figure 1**(e). As seen in the diffusion zone, nitrogen was dissolved in the tempered martensite substrate at 530˚C and nitride precipitates were observed due to the decrease of solubility of nitrogen [12].

Measurement of averaged case thickness was carried out by taking measurements at 5 different positions on the micrograph in **Figure 1**. Images of **Figure 1**(a) to 1(d) showed the thickness of white layers was about 15 ± 5 μm and the diffusion layer thickness was about 300, 550 μm, 800 μm and 1100 μm, respectively. The thickness of diffusion layers was evidently increased with prolonging nitriding duration. The case thickness was strongly correlated to the microhardness profiles of the nitrided specimens as shown in **Figure 2**.

3.2. Microhardness

The cross sectional hardness profiles of nitrided specimens in given conditions are shown in **Figure 2**. All specimens showed high surface microhardness values to about 1100 $HV_{0.1}$. Based on the CNS 14288-G2272 standard [14], the effective case depth of the nitrided specimen was calculated as the case hardness value reaches the substrate hardness plus 50 HV. The effective case depth of the specimens in **Figure 2** as a function of the nitriding duration was shown in **Figure 3**. As can be seen, the effective case depth increased as the increase of the nitriding duration.

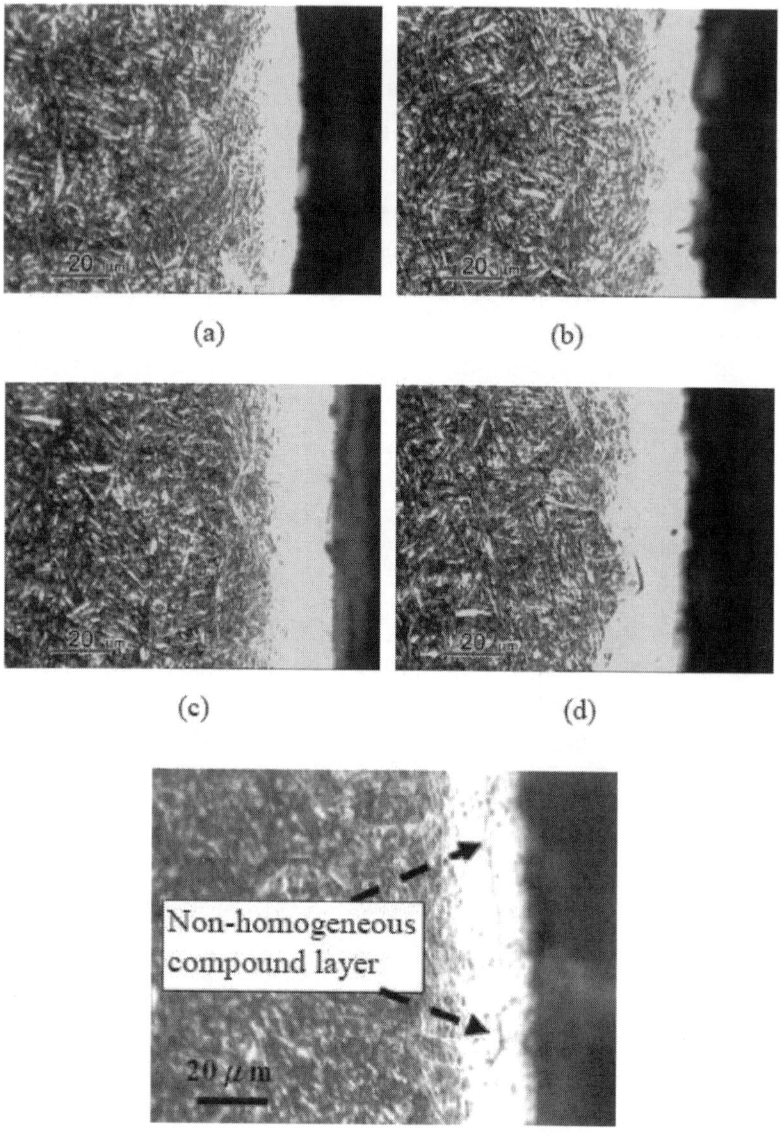

Figure 1. Micrographs of SACM 645 specimens nitrided at given processing conditions, (a) 30C-GN12, (b) 30C-GN24, (c) 30C-GN48, (d) 30C-GN96 and (e) 40C-GN48.

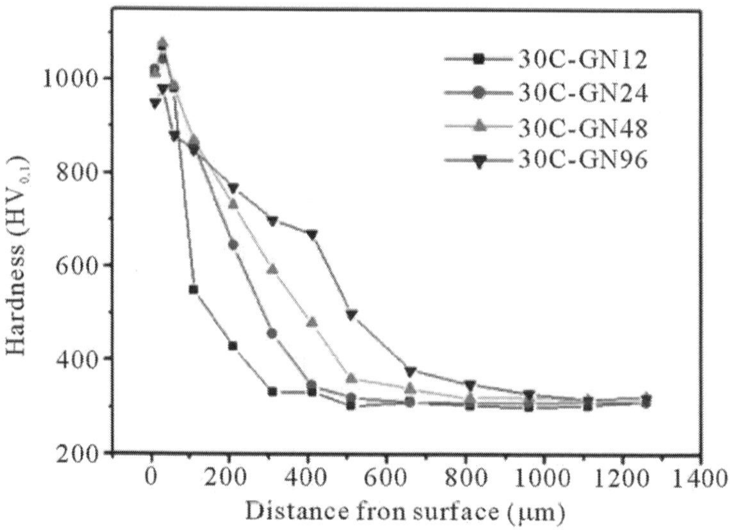

Figure 2. Microhardness profiles of specimens nitrided at given conditions.

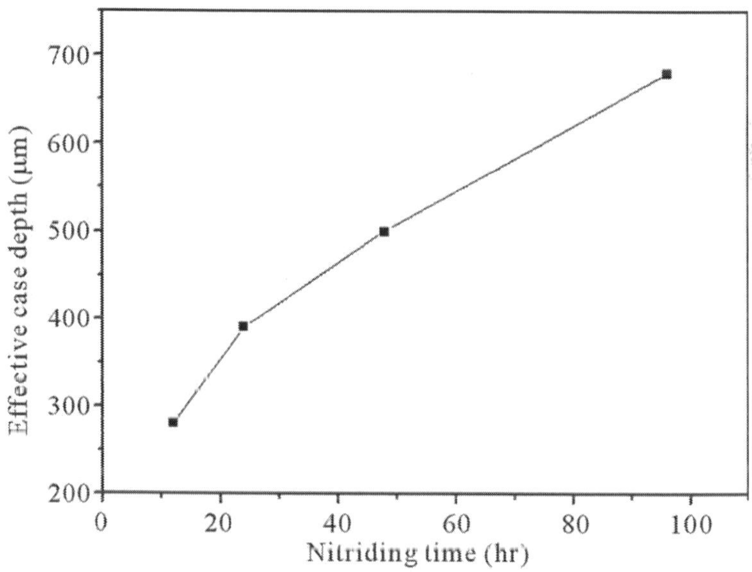

Figure 3. Effective case depth as a function of the nitriding duration.

It can be observed in **Figure 2** that hardness value monotonically drops from the case surface to the substrate. Due to the hardening and trapping effects of the ε-phase (Fe_3N) and γ'-phase (Fe_4N) in the nitrided layer, hardness profiles of the diffusion layer decreased exponentially from the surface. The observed result was in agreement with Suh et al. [2]. A considerable increase in the microhardness of the diffusion zone can be achieved by increasing the amount of supersaturated dissolved nitrogen (solid solution strengthening) [12]. As prolonging the nitriding duration, the effective case depth of specimen treated for 96 h increased to 690 μm.

3.3. XRD Examinations

The X-ray diffraction (XRD) patterns at distinct case depths of 30C-GN48 specimen nitrided at 530°C for 48 h are shown in **Figure 4**. The figure depicted the presence of Fe_2O_3 peaks in the outer case surface. In order to identify the constituents at distinct case depths, the nitrided specimen was carefully sliced off layer by layer of every 30 μm towards the substrate to expose the white compound layer and diffusion layer of the nitrided case.

At case depth of 30 μm, the presence of dominant peaks of the ε-phase (Fe_3N) and minor peaks of the γ'-phase (Fe_4N) were detected. At depth of 110 μm and 200 μm from surface, the presence of broaden peaks of martensite phase showed the trapping effect of nitrogen atoms; it was worth noting that peaks of ε-phase (Fe_3N) and γ'-phase (Fe_4N) were not spotted. At the substrate, the dominant peaks in the XRD pattern were the tempered martensite phase. Other nitrided specimens of different nitriding parameters have depicted similar XRD patterns as those shown in **Figure 4**.

3.4. Wear Behavior

Figure 5(a) shows the effect of nitriding duration on the cumulative wear mass loss of the nitrided specimens. The nitrided specimens exhibited only small amount of cumulative mass losses (in the order of $10^{-2\ g)}$ as a result of the hard compound layers of the nitrided cases. The cumulative mass loss was decreased as the increase of the nitriding duration. The amount of cumulative mass loss of the specimen nitrided for 48 h was only 0.01 g. The wear resistance after nitriding was significantly improved with increase of the nitriding duration. **Figure 5**(b) shows the effect of different substrate hardness values on the cumulative wear mass losses of the nitrided specimens. It was observed that the increases of the substrate hardness values of the specimens nitrided for 48 h resulted in the decreases of the cumulative mass loss.

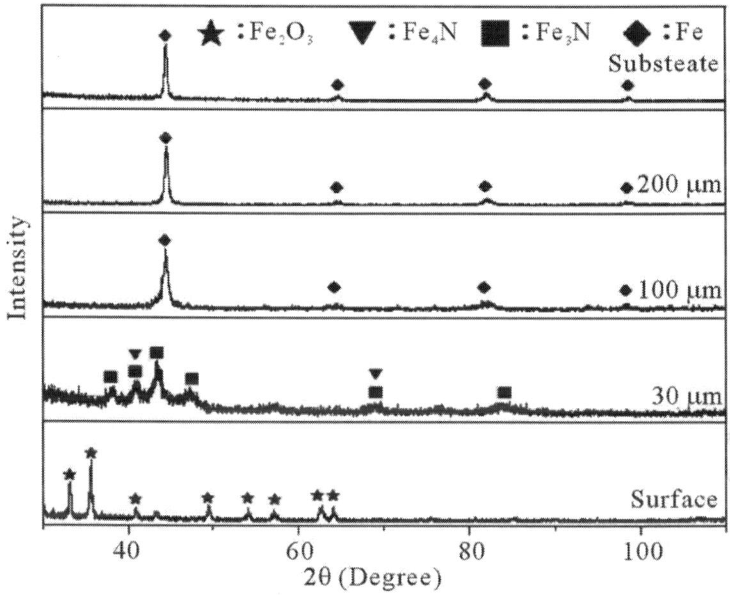

Figure 4. XRD patterns of substrate and gas nitrided layers at 530°C × 48h.

Scanning electron microscope (SEM) micrographs of the wear tracks for both the untreated and nitrided surfaces after wear testing at a load of 95.9 N for 3 h are shown in **Figure 6**. On non-nitrided tempered martensitic specimen (30-TMS), severe abrasive wear was observed (**Figure 6**(a)), whereas the nitrided specimen (30C-GN48) revealed very mild abrasive and adhesive wear tracks on the surface as shown in **Figure 6**(b).

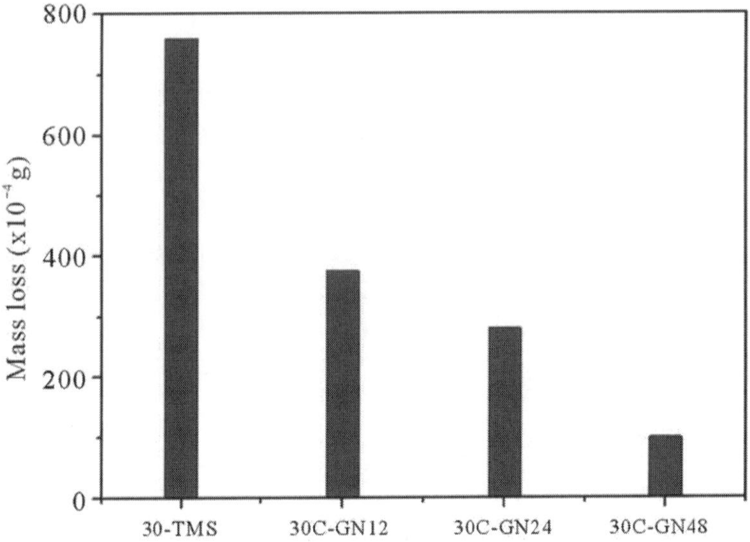

(a)

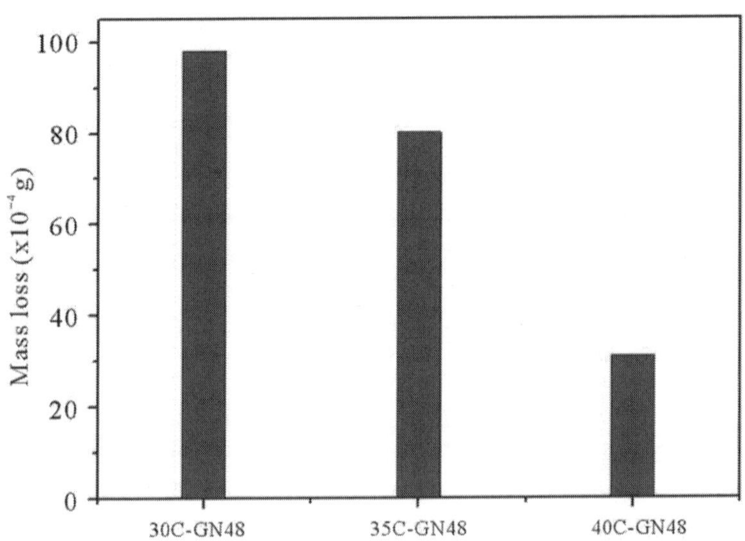

(b)

Figure 5. Mass losses of specimens at given wear test condition for (a) various nitriding durations, and (b) various substrate hardness values.

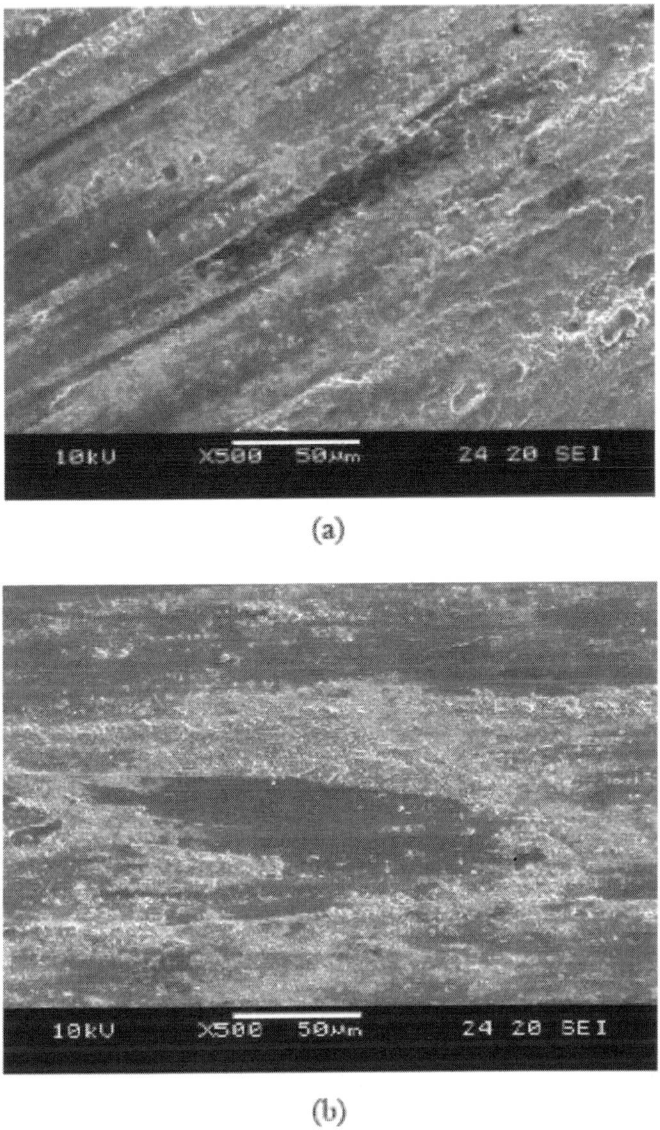

Figure 6. SEM surface morphology of worn specimens undergone tests at 95.9 N for 3 h, (a) 30-TMS, (b) 30C-GN48.

3.5. Fatigue Behavior

Figure 7 shows the fatigue characteristics of untreated and gas nitrided specimens under different nitriding conditions. Fatigue test was carried out in torsion mode at a stress reversal frequency of 3600 rpm. The fatigue limit of the S-N curves for all testing specimens was defined at 1.0 × 10^7cycles without fatigue failure. The fatigue limit of the non-nitrided specimen (30-TMS) was 416 MPa, whereas the fatigue limits of the 30C-GN12, 30C-GN24, 30C-GN48 and 30C-GN96 nitrided specimens were 539, 574, 590 and 600 MPa, respectively. The increase of fatigue limit of the 30C-GN96 specimen was 44.2% compared to the tempered martensitic substrate (30-TMS) specimen observed. Suh et al. [2] reported the rotary bending fatigue characteristics of nitrided SACM 645 steel and found that the fatigue crack was initiated beneath the nitrided layer by residual compressive stress on the surface. They indicated that the crack initiation was hindered by the nitriding layer, while the finding was similar to the work done by Buchhagen and Bell [15]. As shown in previous work [16], the fatigue limit of the specimen with substrate harness value of 35 HRC nitrided for 48 h was reported at a value of 632 MPa, indicating that fatigue limit was affected by the substrate hardness.

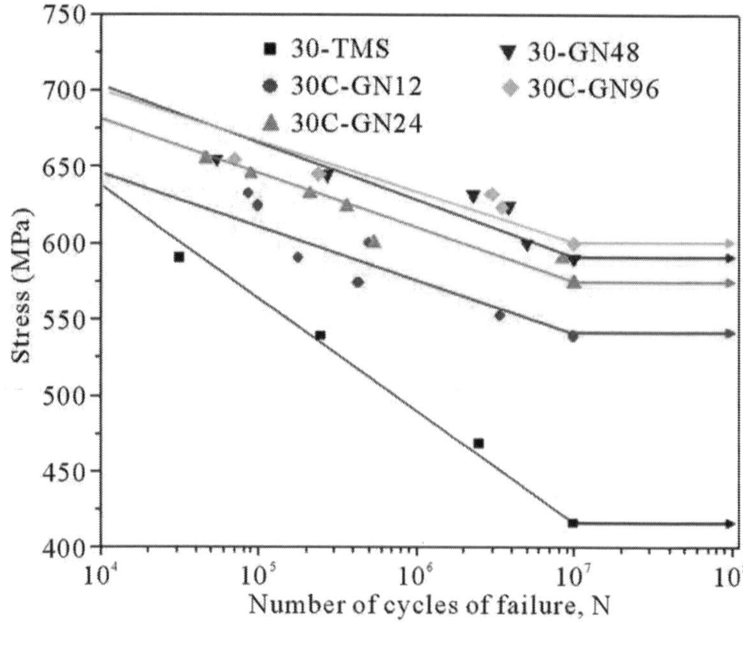

Figure 7. S-N curves of non-nitrided and gas nitrided specimens at given conditions.

Fatigue fracture surfaces of nitrided and non-nitrided specimens after fatigue test were further analyzed using SEM. Fractographic analysis revealed that the fatigue crack in the non-nitrided specimen (30-TMS) was initiated at the surface. A typical surface fatigue initiation site is shown in **Figure 8**(a). In contrast, of the specimens nitrided for 48 h, the fatigue cracks were initiated beneath the nitrided layer close to the substrate-diffusion layer interface as shown in **Figure 8**(b). Similar result was reported by Hussain [17]. As a result, the fatigue limit was improved by gas nitriding, and the fatigue limits increased with increasing nitriding duration.

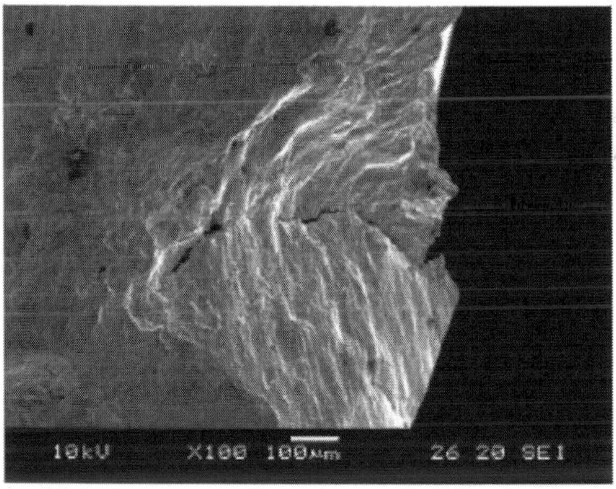

(a)

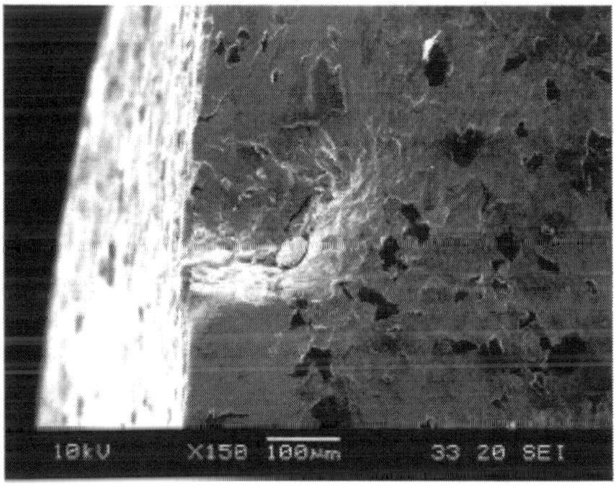

(b)

Figure 8. Fracture surface of fatigue specimen after fatigue test for (a) 30-TMS and (b) 30C-GN48.

3.6. Corrosion Behavior

The polarization curves of non-nitrided and nitrided specimens treated with different nitriding durations are shown in **Figure** 9(a). The polarization curves in **Figure** 9(a) indicated that the non-nitrided specimen did not passivate and endured a continuous dissolution with the corrosion current density (I_{corr}) of 1.0×10^{-5} A/cm2. The corrosion potential (E_{corr}) of the nitrided specimens was shifted to the noble direction as the increases of the nitriding duration. The corrosion current density (I_{corr}) was significantly decreased after nitriding compared to the non-nitrided substrate, which indicated that the nitrided case impeded the anodic dissolution resulted in a very low corrosion current. The corrosion current densities (I_{corr}) of all nitrided specimens were fallen in the range between 3.0×10^{-7} and 6.5×10^{-7} A/cm2.

Figure 9(b) shows the polarization curves of nonnitrided specimen and specimens nitrided for 48 h with different substrate hardness values. A noteworthy shift of the corrosion potentials to more positive values was observed as the substrate hardness of the nitrided specimens decreased. The polarization curves in **Figure** 9(b) depicted that the corrosion potentials (E_{corr}) of 30CGN48, 35C-GN48, 40C-GN48 specimen were about 100 mV, –150 mV and –300 mV. The shift of the corrosion potential (E_{corr}) of nitrided specimens with different substrate hardness may be caused by the non-homogeneous compound layer structure such as cracks in the surface layer. Basso et al. [18] suggested that the porosities or cracks on the compound layer were pathways to connect the 3.5% NaCl solution and substrate through the nitrided case. The corrosion current density (I_{corr}) of the nitrided specimen with substrate hardness of 40 HRC, close to I_{corr} of 30-TMS specimen, was about an order higher than the nitrided specimen with substrate hardness of 30 HRC. Consequently, the corrosion resistance was evidently influenced by the substrate hardness values.

4. CONCLUSIONS

From the X-ray diffraction patterns, Fe_2O_3 peaks were detected on the case surface, and the ε-phase (Fe_3N) and minor peaks of the γ'-phase

(Fe_4N) were appeared at the nitrided case depth of 30 μm. The surface hardness of the nitrided specimens reached values in the range between 1000 $HV_{0.1}$– 1100 $HV_{0.1}$. Wear test data indicated that the mass loss of JIS SACM 645 steel was greatly improved by nitriding processes and the wear resistance of the nitrided specimens was significantly improved by extending nitriding duration.

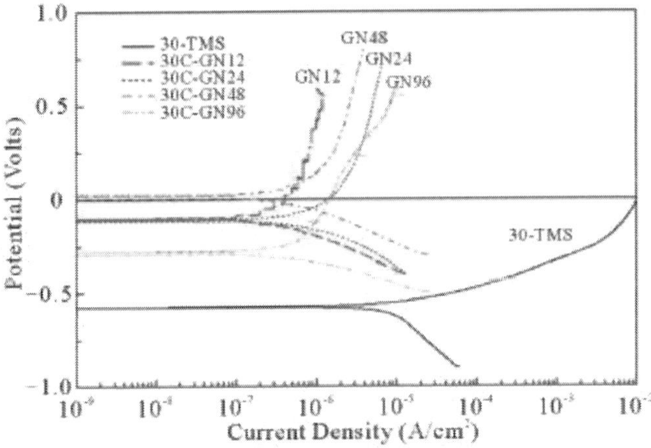

(a)

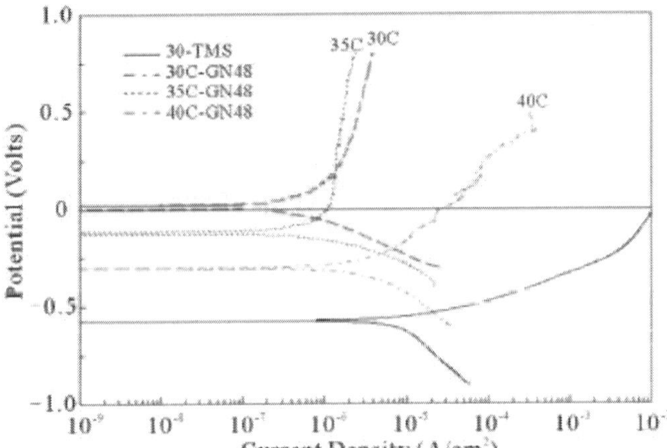

(b)

Figure 9. Polarization curves of non-nitrided and nitrided at 530°C specimens for (a) various nitriding durations and (b) substrate hardness values.

The fatigue limit of the SACM450 steel was greatly improved by nitriding process. It was observed that the fatigue crack of the nitrided specimen was initiated beneath the nitrided layer. With the increase of the diffusion layer thickness, the fatigue limit increased by prolonging nitriding duration. The fatigue limit of 30C-GN96 specimen was 600 MPa which was a 44.2% increase over the 30-TMS specimen in this study.

According to electrochemical measurements, the corrosion potential of nitrided specimen was shifted to the noble direction as the increases of the nitriding duration. The corrosion current densities of all nitrided specimens, in the range of $3.0 \times 10^{-7} \sim 6.5 \times 10^{-7}$ A/cm2, were at least an order lower than that of the non-nitrided substrate, 1.0×10^{-5} A/cm2. As a result, the corrosion resistance was significantly improved by gas nitriding process.

5. REFERENCES

1 C. M. Suh and H. K. Jang, "Effects of Surface Hardening and Residual Stress on the Fatigue Characteristics of Nitrided SACM 645 Steel," International Journal of Modern Physics B, Vol. 17, No. 8-9, 2003, pp. 1633-1639.

2 C. M. Suh, J. K. Hwang, K. S. Son and H. K. Jang, "Fatigue Characteristics of Nitrided SACM 645 According to the Nitriding Condition and Notch," Materials Science and Engineering A, Vol. 392, No. 1-2, 2005, pp. 31-37.

3 J. Baranowskaa and S. E. Franklin, "Characterization of Gas-Nitrided Austenitic Steel with an Amorphous/ Nanocry-Stalline Top Layer," Wear, Vol. 264, No. 9-10, 2008, pp. 899-903.

4 S. Abisset, F. Maury, R. Feurer, M. Ducarroir, M. Nadal and M. Andrieux, "Gas and Plasma Nitriding Pretreatments of Steel Substrates before CVD Growth of Hard Refractory Coatings," Thin Solid Films, Vol. 315, No. 1-2, 1998, pp. 179-185.

5 C. Allen, C. X. Li, T. Bell and Y. Sun, "The Effect of Fretting on the Fatigue Behaviour of Plasma Nitrided Stainless Steels," Wear, Vol. 254, No. 11, 2003, pp. 1106-1112.

6 K. Genel, M. Demirkol and M. Capa, "Effect of Ion Nitriding on Fatigue Behaviour of AISI 4140 Steel," Materials Science and Engineering A, Vol. 279, No. 1-2, 2000, pp. 207-216. doi:10.1016/S0921-5093(99)00689-9

7 N. Limodin and Y. Verreman, "Fatigue Strength Improvement of a 4140 Steel by Gas Nitriding: Influence of Notch Severity," Materials Science and Engineering A, Vols. 435-436, 2006, pp. 460-467. doi:10.1016/j.msea.2006.07.034

8 A. Celik and S. Karadeniz, "Improvement of the Fatigue Strength of AISI 4140 Steel by an Ion Nitriding Process," Surface and Coatings Technology, Vol. 72, No. 3, 1995, pp. 169-173.

9 J. H. Sung, J. H. Kong, D. K. Yoo, H. Y. On, D. J. Lee and H. W. Lee, "Phase Changes of the AISI 430 Ferritic Stainless Steels after High-Temperature Gas Nitriding and Tempering Heat Treatment," Materials Science and Engineering A, Vol. 489, 2008, pp. 38-43.

10 H. W. Lee, J. H. Kong, D. J. Lee, H. Y. On and J. H. Sung, "A Study on High Temperature Gas Nitriding and Tempering Heat Treatment in 17Cr–1Ni–0.5C," Materials & Design, Vol. 30, No. 5, 2009, pp. 1691-1696.

11 D. Q. Peng, T. H. Kim, J. H. Chung and J. K. Park, "Development of Nitride-Layer of AISI 304 Austenitic Stainless Steel during High-Temperature Ammonia Gas-Nitriding," Applied Surface Science, Vol. 256, No. 24, 2010, pp. 7522-7529.

12 F. Ashrafizadeh, "Influence of Plasma and Gas Nitriding on Fatigue Resistance of Plain Carbon (Ck45) Steel," Surface and Coatings Technology, Vols. 174-175, 2003, pp. 1196-1200

13 M. Pellizzari, A. Molinari and G. Straffelini, "Thermal Fatigue Resistance of GAS and Plasma Nitrided 41CrAlMo7 Steel," Materials Science and Engineering, Vol. A352, 2003, pp. 186-194.

14 CNS: 14288-G2272, 2006, 1-6.

15 P. Buchhagen and T. Bell, "Simulation of the Residual Stress Development in the Diffusion Layer of Low Alloy Plasma Nitrided Steels," Computational Materials Science, Vol. 7, No. 1-2, 1996, pp. 228-234.

16 S. H. Yeh, L. H. Chiu and H. Chang, "Influence of Vacuum Heat Treatment an Salt-Bath Nitrocarburization on the Corrosion Behavior of JIS SKD61 and DH31S Hot Work Steels," Advanced Materials Research, Vol. 47-50, 2008, pp. 686-689.

17 K. Hussain, A. Tauqir, A. Haq and A. Q. Khan, "Influence of Gas Nitriding on Fatigue Resistance of Maraging Steel," International journal of fatigue, Vol. 21, No. 2, 1999, pp. 163-168.

18 R. L. O. Basso, R. J. Candal, C. A. Figueroa, D. Wisnivesky and F. Alvarez, "Influence of Microstructure on the Corrosion Behavior of Nitrocarburized AISI H13 Tool Steel Obtained by Pulsed DC Plasma," Surface and Coatings Technology, Vol. 203, No. 10-11, 2009, pp. 1293-1297.

CHAPTER 7

Ultrafine-Grained Austenitic Stainless Steels X4CrNi18-12 and X8CrMnNi19-6-3 Produced by Accumulative Roll Bonding

Mathis Ruppert *, Lisa Patricia Freund, Thomas Wenzl, Heinz Werner Höppel and Mathias Göken

Department of Materials Science and Engineering, Institute I: General Materials Properties, Friedrich-Alexander Universität Erlangen-Nürnberg, Martensstr. 5, 91058 Erlangen, Germany

ABSTRACT

Austenitic stainless steels X4CrNi18-12 and X8CrMnNi19-6-3 were processed by accumulative roll bonding (ARB). Both materials show an extremely high yield strength of 1.25 GPa accompanied by a satisfactory elongation to failure of up to 14% and a positive strain rate sensitivity after two ARB cycles. The strain-hardening rate of the austenitic steels reveals a stabilization of the stress-strain behavior during tensile testing. Especially for X8CrMnNi19-6-3, which has an elevated manganese content of 6.7 wt.%, necking is prevented up to comparatively high plastic strains. Microstructural investigations showed that the microstructure is separated into ultrafine-grained channel like areas and relatively larger grains where pronounced nano-twinning and martensite formation is observed.

KEYWORDS

accumulative roll bonding (ARB); austenitic steel; ultrafine-grained microstructure; strength; nano-twinning; strain rate sensitivity

1. INTRODUCTION

Accumulative roll bonding (ARB) [1] as a process of severe plastic deformation is one of the most effective methods for the production of bulk ultrafine-grained (UFG) materials with a median grain size smaller than 1 μm. The microstructural evolution during ARB is well described in literature, see for example [2,3,4] for details. The mechanical properties of those materials are frequently claimed to be favorable compared to their conventionally grained (CG) counterparts, as a good combination of high strength and satisfactory ductility can be achieved [5,6,7]. The enhanced ductility is often brought into connection with the enhanced strain rate sensitivity [3,6,8,9] of those materials. Moreover, strain rate sensitivity is strongly related to an increased fraction of high angle boundaries [4], as they can act as sources and sinks for dislocations [3]. Although there are numerous publications on ARB available, only a couple of them are dealing with ARB-processing of steel-sheets. Among those, the majority concerns interstitial free bcc steels (IF-steels) with a rather low content of alloying elements, see for example [1,10,11,12]. However, there is very little literature available about ARB of austenitic steels, which will be focused on in the following. Kitahara *et al.* [13] investigated the martensite transformation of an ultrafine-grained Fe-28.5at.-%Ni-alloy with single phase metastable austenite at room temperature. They performed the accumulative roll bonding process with sheets that were pre-heated at 500 °C for 600 s up to 5 cycles. However, each cycle was divided into two passes with a thickness reduction <50%. They achieved an ultrafine-grained microstructure with a mean grain size of 230 nm and a yield strength that was increased by a factor of 4.9 compared to the initial material. Moreover, they showed that the martensite transformation starting temperature decreases with the number of ARB cycles. Another study about ARB of austenitic steel was published by Jafarian *et al.* [14]. They investigated the microstructure and texture development in a Fe-24Ni-0.3C (wt.%) austenitic steel up to 6 ARB-cycles and subsequent annealing. The processing was performed at 600 °C. The texture was found to change from a copper orientation after 1 cycle towards a strong brass component after 6 cycles of ARB. Shen *et al.* [15] performed an accumulative cold rolling process with a thickness reduction of 17% each pass, using sheets of a commercial austenitic stainless steel 304SS, which was pre-heated up to 400 °C prior to each pass. Due to various subsequent heat-treatments, they could achieve materials with different grain sizes. The tensile samples reached a yield strength of up to 1.8 GPa, yet rather small elongation to failure of about 6%. Furthermore twinning and slip of partial dislocations were found to dominate plastic deformation in the ultrafine-grained state. Li *et al.* [16] processed sheets of an austenitic 36%Ni (mass-%) steel up to 6 cycles, with a pre-ARB heat-treatment at 500 °C. They found a rather small grain size of 150 nm and a high misorientation concerning the high angle boundaries.

As no information is available in literature, this work is focused on the strain rate sensitivity of ARB-processed ultrafine-grained austenitic stainless steels. Moreover, the effect of ultrafine grains and pronounced nano-twinning on the mechanical properties of commercially available and technically relevant alloys was addressed.

2. EXPERIMENTAL SECTION

Austenitic steels X4CrNi18-12 (1.4303) and X8CrMnNi19-6-3 (1.4376) were used as initial sheet materials for accumulative roll bonding. Due to the high Ni-content, the austenitic phase of X4CrNi18 12 is stabilized, which leads to a higher cold forming capability. This is desirable, as the material is severely strained during ARB-processing. The austenitic steel X8CrMnNi19-6-3 is a metastable one, which contains 6.7 wt.% of manganese. This induces twinning by plastic deformation, which might lead to satisfactory ductility. The material was delivered by Thyssen Krupp Nirosta GmbH and the chemical composition can be found in Table 1.

Table 1. Chemical compositions of the processed austenitic steels X4CrNi18-12 and X8CrMnNi19-6-3.

Alloy	Composition in wt.%								
	C	Si	Mn	Cr	Ti	P	S	Mo	Ni
X4CrNi18-12	0–0.06	-	0–2.0	17.0–19.0	-	-	-	-	11.0–13.0
X8CrMnNi19-6-3	0.025	0.46	6.76	17.43	0.001	0.029	0.0008	0.23	4.03

The sheets of X4CrNi18-12 and X8CrMnNi19-6-3 had an initial geometry of 25 × 150 × 1 mm (width × length × thickness) and were processed up to three and two cycles of ARB, respectively. Henceforth, the number of performed ARB cycles is denoted by N0-N3. Hereby one cycle of ARB equals a v. Mises equivalent strain of 0.8. During each cycle, the sheets were degreased in acetone and wire brushed with a rotating steel brush. Afterwards, the sheets were pre-heated in an electrical furnace for 5 min at 300 °C and finally roll bonded with a thickness reduction of 50%. After each cycle, edge cracking was cut off and the sheets were prepared accordingly to the scheme described above before the next cycle. In order to determine the mechanical properties of the processed sheets, both Vickers hardness measurements and tensile testing were performed. Therefore, a hardness measurement unit Leco V-100A and an Instron 4505 universal testing machine (Hegewald & Peschke MPT GmbH, Nossen, Germany) for uniaxial tensile testing were utilized. The hardness measurements were conducted at the sheet plane, the rolling plane and the transversal plane. Tensile testing was conducted in rolling direction at room temperature and at strain rates of $10^{-3}s^{-1}$, $10^{-4}s^{-1}$ and $10^{-5}s^{-1}$ in order to determine the strain rate sensitivity. Moreover, microstructural characterization was done using a Zeiss Cross Beam 1540 EsB (Carl Zeiss AG, Oberkochen, Germany) scanning electron microscope in backscattered electron contrast at an acceleration voltage of 11 kV and a working distance of 7–8 mm, as well as a Philips CM 200 transmission electron microscope operated at 200 kV (FEI, Hillsboro, OR, USA).

3. RESULTS AND DISCUSSION

In Figure 1, the results of the hardness measurements in sheet plane, rolling plane and transversal plane are shown for X4CrNi18-12 and for X8CrMnNi19-6-3. Generally an increase of the hardness with the number of ARB cycles can be observed. The largest increase is found for the sheet plane, which is due to the high friction between the sheets and the rolls. This leads to a large shear strain at the surface regions of the sheet [17] and, therefore, to a 15% higher hardness compared to the other planes. Moreover, the hardness of both alloys increases severely during the first cycle of ARB, but only slightly during subsequent cycles. While the hardness of X4CrNi18-12 increases rather constantly between one and three cycles, the hardness of X8CrMnNi19-6-3 already appears to saturate after two ARB cycles.

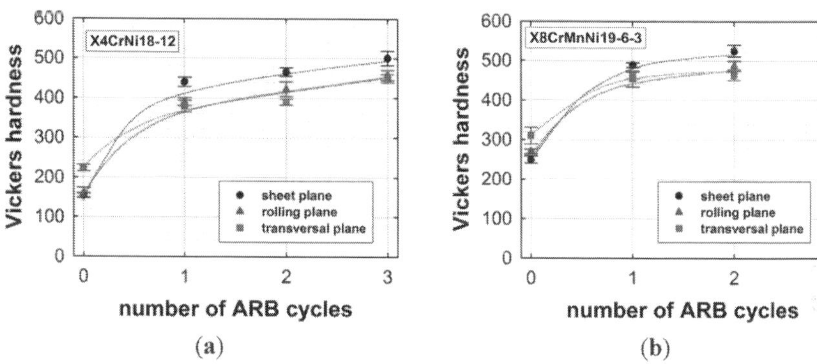

Figure 1. Vickers hardness measurements at the sheet, rolling and transversal planes of the samples after different cycles of ARB for (**a**) X4CrNi18-12 and (**b**) X8CrMnNi19-6-3.

The mechanical performance of both austenitic steels during tensile testing is shown in Figure 2. The ultimate tensile strength (UTS) and the yield strength (YS) increase significantly during the first ARB cycle and the general trend is similar to the hardness measurements. That is to say, both steels reach a YS of around 1.25 GPa after N2 and N3, respectively. The highest increase in strength is found after the first cycle. Therefore, X4CrNi18-12 shows an increase of the UTS by a factor of 1.9 and an increase of the YS by a factor of 3.8. During subsequent ARB-cycles, the strength is further increased, although the relative increase is smaller. The total increase in yield strength compared to the initial material equals a factor of 4.7. The uniform elongation is reduced from 50.4% to 1.2%, while the elongation to failure is reduced from 58% to 7.7%. Similar behavior is found for X8CrMnNi19-6-3, which also shows a strong increase in strength and a strong decrease in ductility after the first ARB cycle. Thereby, the YS increases by a factor of 2.4. The reduction in ductility can be attributed to a decreased hardening rate after severe plastic deformation. Nevertheless, both materials show an excellent combination of strength and ductility. The ARB-processed X8CrMnNi19-6-3 especially performs very well and reaches an UTS between 1.3 and 1.5 GPa, while it maintains an elongation to failure between 10 and 14%. Generally, the shape of the stress-strain curves changes after the first ARB cycle. While the N0 samples deform mainly uniformly during the tensile testing, the ARB-processed ones start necking at pretty low strains but show long post-necking deformation. This transition of tensile deformation behaviors in ultrafine-grained materials was discussed in detail for aluminum by Yu *et al.* [18]. Referring to this publication, tensile stress-strain curves can be categorized into four different characteristic types, in dependence of the grain size and the testing temperature. The N0-curves obtained in the present study a can be assigned to Type IV, which means that the curve shows continuous strain-hardening. This is typically observed in coarse-grained materials with a grain size larger than 4 μm. The stress-strain curves of the ARB processed X4CrNi18-12 can be assigned to Type II, which means that the curves exhibit a distinct yielding peak followed by strain-softening. Type II behavior is observed for grain sizes between 0.4 μm and 1 μm. According to Yu *et al.* [18], the yielding peak is brought into connection with a lack of mobile dislocations, due to the large dislocation sink area provided by grain boundaries. This so-called yield-drop was also found for ultrafine-grained aluminum AA1100 and IF-steel [19], UFG Cu [20], UFG Ti [21] and also for cold-rolled high-manganese austenitic steel [22], which shows twinning induced plasticity. Concerning X8CrMnNi19-6-3, also a distinct yield point followed by strain softening can be observed. In Figure 2d, the true strain hardening rate is determined between the uniform elongation and the elongation to failure and plotted over the true plastic strain for both austenitic steels after N2. The curves were obtained by calculating the true stress strain curves (Figure 2c) from the engineering data and by determining the derivative of those curves. It has to be considered, that the actual cross-section of the tensile samples during tensile testing was not measured. However, the strain hardening curves can be compared qualitatively. It can be observed, that both materials initially show the same behavior up to 5% of plastic strain. That is to say, the strain hardening rate is reduced between the yield point and 3% of true plastic strain. Up to 5% the hardening rate for both alloys is increasing again. Afterwards, the strain-hardening rate of X4CrNi18-12 drastically decreases, while for X8CrMnNi19-6-3 the strain hardening rate increases until it suddenly drops as soon as the sample breaks. This is most likely due to failure because of ARB-related bonding defects. On the one hand, the stabilization of the stress-strain curves could be influenced by an enhanced strain rate sensitivity, which is typically observed for ultrafine-grained fcc metals. Thereby, thermally activated annihilation of dislocations is assumed to play a decisive roll and might lead to increased post-necking strains. On the other hand, both steels show a three-stage work hardening behavior, which is typical for materials with pronounced twinning activity and which is also found for high manganese TWIP (twinning induced plasticity) steels. It appears that the second stage is more distinct for X8CrMnNi19-6-3 compared to X4CrNi18-12. This might be due to the elevated manganese content, which leads to a higher twinning activity during tensile deformation that could stabilize the deformation behavior. The decreasing hardening rate after 5% of plastic strain, which is observed for X4CrNi18-12, might be attributed to a saturation in the amount of twinned grains. This saturation is not reached for X8CrMnNi19-6-3 during plastic straining. However, also martensitic transformation might contribute to the stabilization of the stress-strain curves. To gain more insight, the strain rate sensitivity and the twinning behavior of the materials were investigated.

The strain rate sensitivity (SRS) of the austenitic steels was determined from tensile testing experiments at strain rates of $10^{-3}s^{-1}$, $10^{-4}s^{-1}$ and $10^{-5}s^{-1}$. SRS has to be determined under constant microstructural conditions. Thus, the determination of the SRS from tensile tests is rather crucial, as a microstructural stable condition is hardly achieved. In order to minimize this problem, all stress-strain curves were analyzed at maximum stress, which appears to be a good compromise between the evolution of the microstructure and the limitations of the onset of necking. Therefore, true stress–true strain diagrams were plotted and the maximum stress was evaluated from the different curves. Afterwards, the determined values where plotted over the corresponding strain rate according to [23,24]. Figure 3 reveals that in the initial N0 condition X4CrNi18-12 shows a small positive SRS of around 0.007 at room temperature, while the SRS of X8CrMnNi19-6-3 is around zero. After the first and the second ARB cycle, the SRS of X4CrNi18-12 is slightly decreased to around zero. However, it increases distinctly during the third ARB cycle to 0.017. A similar trend can be found for X8CrMnNi19-6-3, where the SRS remains around zero after the first cycle, but increases after the second cycle to 0.021. The SRS after two cycles in the case of X8CrMnNi19-6-3 and after three cycles in the case of X4CrNi18-12 are in the range of SRS found for other ultrafine-grained fcc materials in literature [6,9,25]. Pronounced strain rate sensitivity is frequently brought into connection with an increased fraction of high angle boundaries, which is typically found in ultrafine-grained fcc metals, see for example [4]. Those high angle grain boundaries can act as sources and sinks for dislocations [3], which are able to stabilize the stress-strain behavior and eventually lead to a higher elongation to failure [6,25]. Consequently, Figure 2 reveals a higher elongation to failure for X8CrMnNi19-6-3 compared to X4CrNi18-12, as the strain rate sensitivity was determined to be higher. However, the SRS usually increases with the number of ARB cycles. Nevertheless, the SRS of X4CrNi18-12 is decreased after the first ARB cycle, which is a rather untypical behavior. In order to clarify more about this point, the microstructures of both alloys were investigated in detail by means of SEM and TEM.

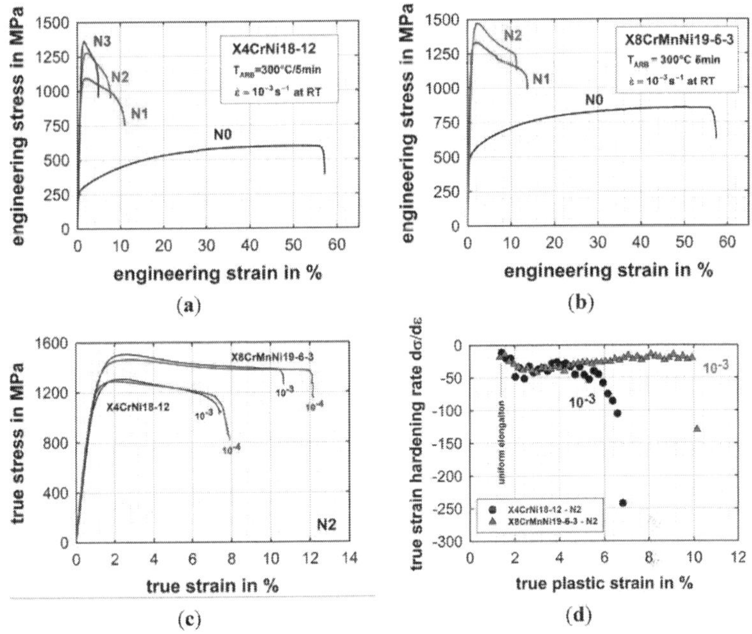

Figure 2. Engineering stress-strain curves at RT and a strain rate of $10^{-3}s^{-1}$ for **(a)** X4CrNi18-12 and **(b)** X8CrMnNi19-6-3. **(c)** Comparative plot of true stress-strain curves of X4CrNi18-12 and X8CrMnNi19-6-3 at strain rates of $10^{-3}s^{-1}$ and $10^{-4}s^{-1}$. Stress-strain curves for $10^{-5}s^{-1}$ are omitted for reasons of clarity. **(d)** True strain hardening rate for X4CrNi18-12 and X8CrMnNi19-6-3 during tensile testing at a strain rate of $10^{-3}s^{-1}$ after N2 cycles of ARB between the uniform elongation and the elongation to failure.

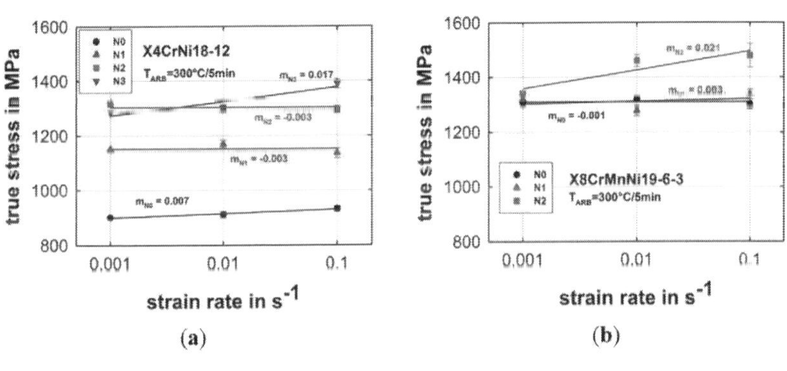

Figure 3. True maximum stress in MPa *vs.* strain rate for **(a)** X4CrNi18-12 and **(b)** X8CrMnNi19-6-3. The slope of the curves equals the strain rate sensitivity m_{Nx}.

Concerning the stabilized austenitic steel X4CrNi18-12 in the initial N0 condition, which can be seen in Figure 4a, the grain size of the equiaxial austenite grains is found to be in the range between 15–30 μm. Moreover, various annealing twins can be observed. The coarse grain structure leads to a high strain hardening capability and therefore to a high uniform elongation. Figure 4b reveals that after one ARB-cycle (ε = 0.8) the grains appear highly deformed and pronounced mechanical twinning and martensite formation are clearly visible. Generally, the microstructure shows various types of microstructural features. On the one hand, large grains with a grain size clearly above 1 μm. These grains are filled up by nano-twins and martensite needles. On the other hand, shear band like regions were found, indicating highly localized plastic deformation with very small grain sizes in the UFG regime. With increasing number of ARB cycles, the fraction of the UFG area is extended and the grain size is further reduced far below 1 μm, see Figure 4c. Moreover, the heterogeneity of the microstructure related to the above described features decreases. Nevertheless, some of the ultrafine grains still contain nano-sized twins in the range of 5–20 nm, which could be observed in the TEM, see Figure 4d. According to Mangonon *et al.*[26], martensitic transformations in Fe-Cr-Ni alloys proceed in the sequence γ (fcc)→ε (hcp)→α' (tetragonal-bcc). Thereby, α' is preferably built at intersections of two ε bands and near regions where ε bands are adjacent to twins or grain boundaries. Furthermore, α' first appears to be needle-shaped and becomes more lath-shaped during subsequent growth. Hereby, the growth of α' leads to a reduction of the ε phase. Shen *et al.* [27] found that both ε and twins act as an intermediate phase during the transformation form γ-austenite to α'-martensite. Above a certain strain level, they observed that the twin density is decreasing, while the martensite density is further increasing. Huang *et al.* [28] performed detailed TEM-investigations on the nucleation mechanism of deformation-induced martensite in an austenitic steel under ECAP-deformation. They observed similar behavior as described above; however they could also find α' nucleating at the intersection of two deformation twins and micro shear-bands. Accordingly, the diffraction patterns of those zones become rather complex (compare inset of Figure 5d).

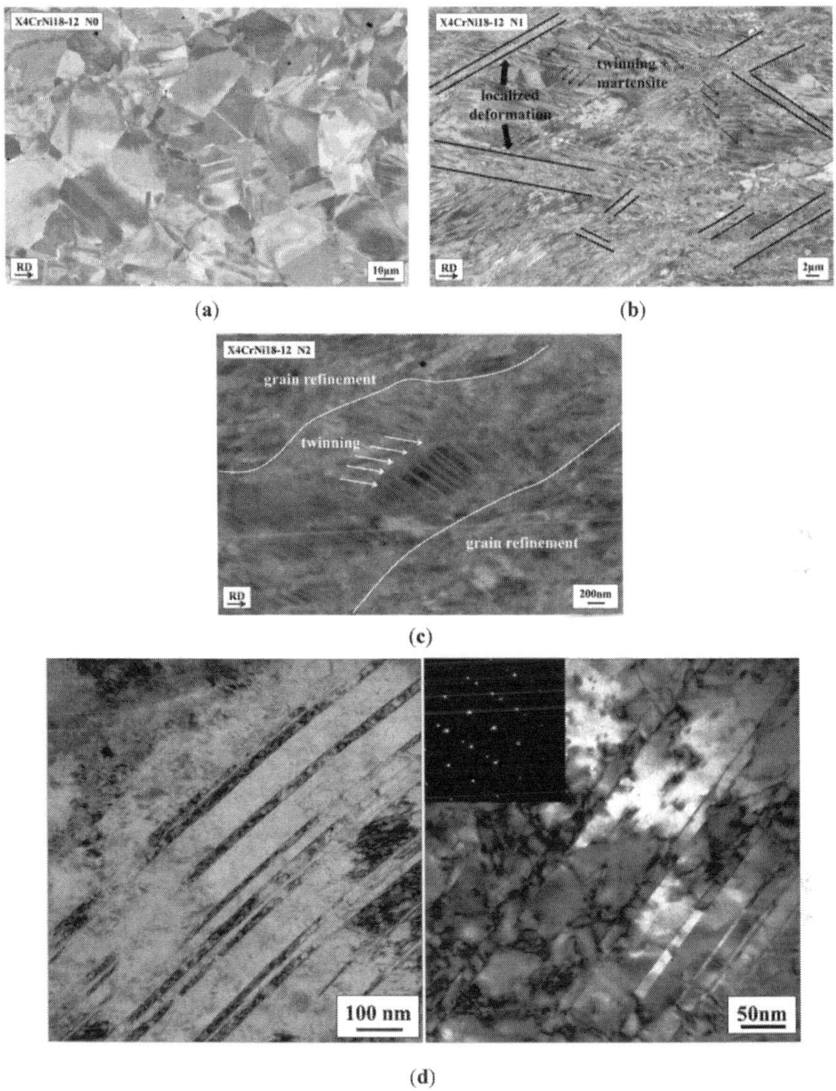

Figure 4. SEM micrographs of ARB-processed austenitic steel X4CrNi18-12 after (**a**) N0, (**b**) N1 and (**c**) N2. The arrows indicate areas where twins but also martensite were found. In (**d**) TEM captions of a highly twinned region with the corresponding diffraction pattern is shown after N2 cycles of ARB.

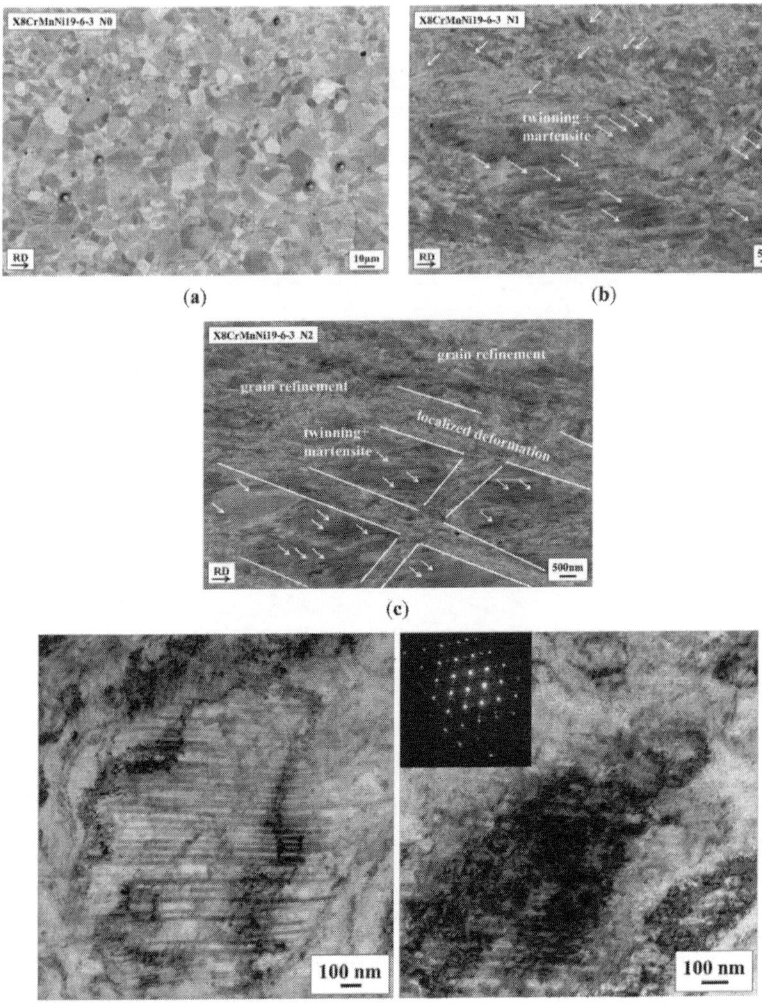

(a) (b)

(c)

(d)

Figure 5. SEM micrographs of ARB-processed austenitic steel X8CrMnNi19-6-3 after (**a**) N0, (**b**) N1 and (**c**) N2 cycles. The arrows indicate areas where twins but also martensite was found. (**d**) TEM bright-field caption after N2 ARB-cycles and the corresponding diffraction pattern of this area.

Generally, the microstructural evolution of the austenitic steel X8CrMnNi19-6-3 is quite similar to that of X4CrNi18-12. That is to say, X8CrMnNi19-6-3 also shows equiaxed coarse grains in the initial N0 condition. The grain size is about 10 μm, which is slightly smaller than that of X4CrNi18-12 and there is also a smaller amount of annealing twins, see Figure 5a. After the first ARB cycle, the microstructure already appears highly deformed, and both twinning and martensite are found in the whole microstructure, which can be seen in Figure 5b. Furthermore, some larger grains in the range between 0.5 and 1.5 μm are found, which are rather randomly surrounded by very small grains. After the second ARB cycle, oval-shaped areas with a diameter of 2–5 μm are forming, see Figure 5c. Those areas are divided by zones of localized plastic deformation and very small grain sizes, which appear to be shear bands orientated 45° to rolling direction. Moreover, the fraction of ultrafine-grained areas is clearly increased compared to the N1 condition. In TEM investigations γ-austenite, martensitic areas as well as twinning were found. Accordingly, in [28], the observed diffraction patterns were quite complex. In Figure 5d the diffraction pattern of such an area is representatively shown. Although indications for twinning and martensite could be found, the heavily deformed structures did not allow clear identification and separation of the different phases.

After the first ARB cycle, the microstructure of both alloys is dominated by larger grains containing mechanical twins and martensitic needles. As the fraction of ultrafine grains is rather small, twinning and martensitic transformation can be regarded as the dominating deformation mechanisms. Thus, the highly deformed microstructure contributes to the pronounced increase in strength. Twinning is also assumed to significantly contribute to plastic deformation during tensile testing, which stabilizes the stress-strain behavior. As a consequence of that, elongation to failure becomes relatively high, see Figure 2. Because of pronounced twinning after one ARB cycle and reduced interaction of dislocations with grain boundaries the strain rate sensitivity of both austenitic steels is approaching zero after the first ARB cycle, as shown inFigure 3. This appears to be in contrast to observations by Lu *et al.* [29], who report a positive effect of twins on the SRS in UFG Cu. Nevertheless, both the grain size and the twin thickness are in a much smaller regime in that study. Moreover. martensite formation found in this work might affect the SRS. As the fraction of ultrafine-grained regions is increasing again during subsequent ARB cycles and the twinned areas are reducing, the dislocation grain boundary interaction is enhanced and the strain rate sensitivity is also increasing again. This leads to a positive strain rate sensitivity, which is usually observed in many fcc metals. Mechanisms based on the thermally activated interaction of dislocations with grain boundaries, which might provide a good explanation for the observed increase in SRS, are given by Blum *et al.* [3] in terms of thermally activated annihilation of dislocations at grain boundaries or by Kato *et al.* [30] in terms of thermally activated dislocation depinning at grain boundaries. Moreover, grain boundary sliding cannot be completely neglected and might also contribute to the enhanced SRS to a certain extent. However, when ductility is regarded, the situation becomes more complex: On the one hand, with increasing number of ARB cycles, decreasing elongation to failure is observed as the twinning capability is further decreased and the fraction of martensitic areas becomes more pronounced. On the other hand, the SRS, which is known to positively affect ductility, see Figure 2c, is increased for both alloys. Consequently, the elongation to failure is still rather satisfactory. Therefore, it is assumed that the increased strain rate sensitivity, martensitic transformation as well as twinning appear to contribute to the good ductility during tensile testing.

4. CONCLUSIONS

In the present study, commercially available austenitic steels X4CrNi18-12 and X8CrMnNi19-6-3 were processed by accumulative roll bonding at elevated temperatures. For both alloys, pronounced twinning and needles like martensite were found after the first ARB cycle. An increase in yield strength by a factor of 3.9 and 2.4, respectively, was found. During subsequent ARB cycles, the fraction of ultrafine-grained microstructure was clearly increased and a very high yield strength of 1.25 GPa was achieved. For X8CrMnNi19-6-3, the high strength was combined with a satisfactory ductility of more than 10%. The ultrafine-grained regions appeared in channels, dividing coarser areas where pronounced twinning with twin size of 5–20 nm was found. Typical for ultrafine-grained fcc metals, positive strain rate sensitivity between 0.017 and 0.021 was revealed for both alloys, as soon as the fraction of ultrafine-grained regimes was increased. Strain rate sensitivity, twinning and martensitic transformation appear to contribute to the satisfactory ductility.

ACKNOWLEDGMENTS

The authors would like to thank the German Research Council (DFG) for their financial support and the Cluster of Excellence "Engineering of Advanced Materials", which is funded within the framework of its "Excellence Initiative".

AUTHOR CONTRIBUTIONS

Mathis Ruppert and Thomas Wenzl performed the mechanical characterization and the SEM investigations. Lisa Patricia Freund prepared the TEM images. Heinz Werner Höppel and Mathias Göken supervised the work and discussed the results with the other authors.

CONFLICTS OF INTEREST

The authors declare no conflict of interest.

REFERENCES

1 Saito, Y., Tsuji, N.; Utsunomiya, H.; Sakai, T.; Hong, R.G. Ultra-fine grained bulk aluminum produced by accumulative roll-bonding (ARB) process. *Scr. Mater.* **1998**, *39*, 1221–1227.

2 Huang, X.; Tsuji, N.; Hansen, N.; Minamino, Y. Microstructural evolution during accumulative roll-bonding of commercial pure aluminum. *Mater. Sci. Eng. A* **2003**, *340*, 265–271.

3 Blum, W.; Zeng, X.H. A simple dislocation model of deformation resistance of ultrafine-grained materials explaining Hall-Petch

strengthening and enhanced strain rate sensitivity. *Acta Mater.* **2009**, *57*, 1966–1974.

4 Hughes, D.A.; Hansen, N. High angle boundaries formed by grain subdivision mechanisms. *Acta Mater.* **1997**, *45*, 3871–3886.

5 Saito, Y.; Utsunomiya, H.; Tsuji, N.; Sakai, T. Novel ultra-high straining process for bulk materials—Development of the accumulative roll bonding (ARB) process. *Acta Mater.* **1999**, *47*, 579–583.

6 Höppel, H.W.; May, J.; Göken, M. Enhanced strength and ductility in ultrafine-grained aluminium produced by accumulative roll bonding. *Adv. Eng. Mater.* **2004**, *6*, 781–784.

7 Valiev, R.Z.; Alexandrov, I.V.; Zhu, Y.T.; Lowe, T.C. Paradox of strength and ductility in metals processed by severe plastic deformation. *J. Mater. Res.* **2002**, *17*, 5–8.

8 Wei, Q. Strain rate effects in the ultrafine grain and nanocrystalline regimes-influence on some responses. *J. Mater. Sci.***2007**, *42*, 1709–1727.

9 May, J.; Höppel, H.W.; Göken, M. Strain rate sensitivity of ultrafine-grained aluminium processed by severe plastic deformation. *Scr. Mater.* **2005**, *53*, 189–194.

10 Tsuji, N.; Ueji, Y.; Minamino, Y. Nanoscale crystallographic analysis of ultrafine grained IF steel fabricated by ARB process. *Scr. Mater.* **2002**, *47*, 69–76.

11 Kamikawa, N.; Sakai, T.; Tsuji, N. Effect of redundant shear strain on microstructure and texture evolution during accumulative roll-bonding in ultralow carbon IF steel. *Acta Mater.* **2007**, *55*, 5873–5888.

12 Lee, S.-H.; Saito, Y.; Park, K.-T.; Shin, D.H. Microstructures and mechanical properties of ultra low carbon IF steel processed by accumulative roll bonding process. *Mater. Trans.* **2002**, *43*, 2320–2325.

13 Kitahara, H.; Tsuji, N.; Minamino, Y. Martensite transformation from ultrafine grained austenite in Fe-28.5 at. % Ni.*Mater. Sci. Eng. A* **2006**, *438–440*, 233–236.

14 Jafarian, H.; Eivani, A. Texture development and microstructure eolution in metastable austenitic steel processed by accumulative roll bonding and subsequent annealing. *J. Mater. Sci.* **2014**, *49*, 6570–6578.

15 Shen, Y.F.; Zhao, X.M.; Sun, X.; Wang, Y.D.; Zuo, L. Ultrahigh strength of ultrafine grained austenitic stainless steel induced by accumulative rolling and annealing. *Scr. Mater.* **2014**.

16 Li, B.; Tsuji, N.; Minamino, Y. Microstructural evolution in 36%Ni austenitic steel during the ARB process. *Mater. Sci. Forum* **2006**, *512*, 73–78.

17 Lee, S.-H.; Saito, Y.; Tsuji, N.; Utsunomiya, H.; Sakai, T. Role of shear strain in ultragrain refinement by accumulative roll-bonding (ARB) process. *Scr. Mater.* **2002**, *46*, 281–285.

18 Yu, C.Y.; Kao, P.W.; Chang, C.P. Transition of tensile deformation behaviors in ultrafine-grained aluminum. *Acta Mater.* **2005**, *53*, 4019–4028.

19 Tsuji, N.; Ito, Y.; Saito, Y.; Minamino, Y. Strength and ductility of ultrafine grained aluminum and iron produced by ARB and annealing. *Scr. Mater.* **2002**, *47*, 893–899.

20 An, X.H.; Wu, S.D.; Zhang, Z.F.; Figueiredo, R.B.; Gao, N.; Langdon, T.G. Enhanced strength-ductility synergy in nanostructured Cu and Cu-Al alloys processed by high-pressure torsion and subsequent annealing. *Scr. Mater.* **2012**,*66*, 227–230.

21 Li, Z.; Fu, L.; Fu, B.; Shan, A. Yield point elongation in fine-grained titanium. *Mater. Lett.* **2013**, *96*, 1–4.

22 Saha, R.; Ueji, R.; Tsuji, N. Fully recrystallized nanostructure fabricated without severe plastic deformation in high-Mn austenitic steel. *Scr. Mater.* **2013**, *68*, 813–816.

23 Hart, E.W. Theory of the tensile test. *Acta Metall.* **1967**, *15*, 351–355.

24 Ghosh, A.K. On the measurement of strain-rate sensitivity for deformation mechanism in conventional and ultra-fine grain alloys. *Mater. Sci. Eng. A* **2007**, *463*, 36–40.

25 Höppel, H.W.; May, J.; Eisenlohr, P.; Göken, M. Strain-rate sensitivity of ultrafine-grained materials. *Zeitschrift für Metallkunde* **2005**, *96*, 566–571.

26 Mangonon, P.; Thomas, G. The martensite phases in 304 stainless steel. *Metall. Trans.* **1970**, *1*, 1577–1586.

27 Shen, Y.F.; Li, X.X.; Sun, X.; Wang, Y.D.; Zuo, L. Twinning and martensite in a 304 austenitic stainless steel. *Mater. Sci. Eng. A* **2012**, *552*, 514–522.

28 Huang, C.X.; Yang, G.; Gao, Y.L.; Wu, S.D.; Li, S.X. Investigation on the nucleation mechanism of deformation-induced martensite in an austenitic stainless steel under severe plastic deformation. *J. Mater. Res.* **2007**, *22*, 724–729.

29 Lu, K.; Lu, L.; Suresh, S. Strengthening materials by engineering coherent internal boundaries at the nanoscale. *Science***2009**, *324*, 349–352.

30 Kato, M. Thermally activated dislocation depinning at a grain boundary in nanocrystalline and ultrafine-grained materials. *Mater. Sci. Eng. A* **2009**, *516*, 276–282.

CHAPTER 8

Effect of V-Ti on the Microstructure and Abrasive Wear Behavior of 6CrC Cast Steel Mill Balls

S. I. Maldonado-Ruiz[1*], P. Orozco-González[1], V. H. Baltazar-Hernández[1], A. Bedolla-Jacuinde[2], M. A. L. Hernández-Rodríguez[3]

[1]Unidad Académica de Ingeniería, Universidad Autónoma de Zacatecas, Zacatecas, México

[2]Instituto de Investigaciones Metalúrgicas, Universidad Michoacana, Morelia, México

[3]Universidad Autónoma de Nuevo León, Monterrey, México

ABSTRACT

Iron-chromium cast alloys are basically abrasive wear resistant materials particularly employed in mining industry; these alloys are often utilized in the manufacture of milling balls. In particular, high Cr and high C cast alloys have been subjected of significant research; for instance, most reports have been addressed on analyzing the relation between microstructure and the abrasive wear behavior; however, there exist a reduced number of reports on relatively low Cr and low C cast alloys. In this research, five low Cr cast steels containing additions of V and Ti were melted in an open atmosphere induction furnace. Comparisons on the morphology, size, type and distribution of carbides were carried through optical microscopy, SEM and XRD. Hardness testing was employed at room temperature with the purpose of correlate to wear behavior. A laboratory pilotplant ball-mill set with a batch of ore was utilized in order to evaluate the abrasive wear resistance. According to microstructure observations, a martensitic primary matrix was revealed in all specimens. The fraction of M_7C_3 and M_3C interdendritic eutectic carbides varied according to alloying level. Further results indicated that variations in the shape and size of M_7C_3 and M_3C along with the presence of V and Ti carbides influenced on the abrasive wear behavior of low Cr cast steel mill balls.

KEYWORDS

Wear-Resistance, Grinding-Media, Cr-C Cast-Steel

1. INTRODUCTION

White cast iron is widely utilized in applications that require resistance to wear [1] [2]. White cast iron is extensively used in ore processing, mining, metallurgy and energy industries [3] -[6] . In particular, iron-chromium white cast irons (C-Cr-Fe) are specific groups of materials characterized as highly resistant to abrasive wear [7] [8] . Hence, there is a type of iron-chromium cast alloys which are often utilized in the manufacture of milling balls. Owe to milling ball application in ore grinding; the ball surface is exposed to different abrasive mediums like: silica (quartz), corundum, feldspar, basalts and other mineral particles [1].

Iron chromium cast irons are based on the C-Cr-Fe diagram as it shows the phase relationships in the Fe-rich corner [9]. It is worth noting that these cast irons are comprised of iron-chromium carbides [10] (i.e. M_3C, M_7C_3) embedded into an iron matrix [6] [11] -[13]; such carbides are favored with presence of high carbon content [9] . Figure 1 shows the Fe-rich corner of the metastable C-Cr-Fe liquidous diagram [9] along with superimposed areas (shadow regions); it is important to mention that highlighted areas indicate these C-Cr-Fe alloys are being subject to extensive research [7] [14] .

In most cases, iron-chromium cast irons are comprised of high carbon content and high chromium along with small amounts of molybdenum, nickel, vanadium, titanium, cupper, niobium, tungsten, and boron [15] . It has been reported that high chromium and carbon cast irons are extremely abrasion resistant owe to their carbide content and their orientation with respect to the wear surface played an important role in the wear behavior [16] [17] . A primary network of M_7C_3surrounded by an austenitic matrix and eutectic carbides, influenced on the wear mechanism in high C and Cr cast alloys [18] . Further, improved wear resistance has been reported if adding Mo, V and Ti, which in turn, has been related to the fine distribution of the eutectic aggregates [19] . Although there are a large number of reports regarding to abrasive wear of high-chromium cast irons; few reports are encountered having reduced amount of C and Cr when combined with a varied elements such V-Ti and their effects on the abrasive wear resistance.

This work aims on evaluating the microstructure and the abrasive wear resistance of low chromium cast steel containing varied amounts of V and Ti by employing a small-scale pilot-plant mill.

2. EXPERIMENTAL PROCEDURE

Five different selected cast alloys were used in this research. The selected cast alloys resulted from analysis of previous studies [18] [19] as well as from a split-plot experimental design of trial [20] . The whole batch of cast alloys were melted in an open atmosphere induction furnace having a hollow capacity of 15 kg per batch, and poured into a sand mold.

The chemical composition of all alloys is listed in Table 1; it can be noted, that all alloys resulted in relatively low Cr and C content; as a matter of fact, the alloys studied in this work are located near the left side corner of the C-Cr-Fe-rich corner diagram (i.e. ~1.5% C and ~5% Cr), indicated by numbers in Figure 1. In particular, samples numbered as 2-5 resulted in variations of approximately from 2 to 4 pct in Mn, 0.5 to 1.0 pct of Mo, and small variations of 0.25 to 0.5 pct of Ti and V. Other chemical elements were kept approximately constant at about: 1.5 pct C, 5 pct Cr, 0.3 pct Si and 1 pct Cu (Table 1). V and Ti were missing in

Sample 1 with the purpose of benchmark the lean chemistry alloy (i.e. Cr-Mo alloy).

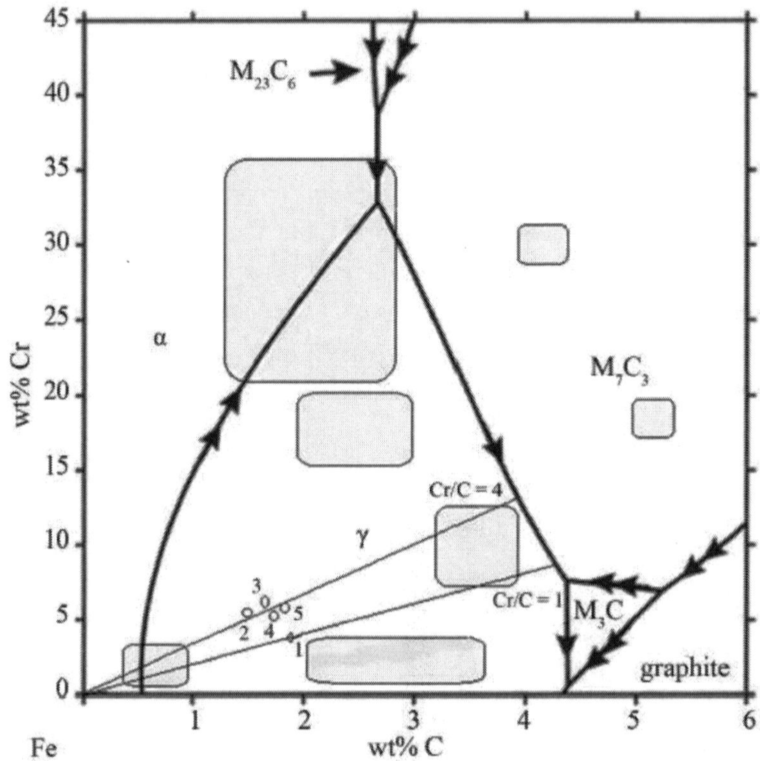

Figure 1. C-Cr-Fe-rich corner diagram, (the sign indicates the selected analyzed in this work, along with previous studies).

Optical microscopy was employed to analyze the microstructure. Beraha's solution was used for etching all specimens. The volume fraction of the carbides was determined using quantitative metallographic by image analyzer. Samples were analyzed by scanning electron microscopy (SEM) and X-ray diffraction (XRD). Rockwell C hardness testing and microhardness HV (300 g @ 15 s) were applied in the as-cast condition.

Table 1. Chemical composition of the cast alloys.

Sample No.	Weight percent									
	C	Mn	Si	Cr	Mo	V	Ti	Cu	S	P
1	1.866	3.596	0.252	4.050	1.210	-	-	0.937	0.048	0.031
2	1.507	3.593	0.252	5.741	0.710	0.224	0.234	0.932	0.040	0.041
3	1.65	1.536	0.503	7.974	0.182	0.475	0.523	0.839	0.035	0.043
4	1.735	3.458	0.252	5.057	1.234	0.582	0.561	0.924	0.052	0.022
5	1.85	1.897	0.301	5.883	0.300	0.271	0.267	0.834	0.040	0.052

Small-scale pilot-plants have been recently introduced in order to perform abrasive wear testing on mill balls, such as those employed in industrial plants. The advantage of using a small-scale pilot-plant ball mill for testing materials over standard wear tests is that some parameters (related to industrial application) can be simulated under laboratory conditions and can be directly correlated to ball consumption [21] .

A pilot-plant ball-mill having a total capacity of 10 l was utilized in this research in order to evaluate the abrasive wear behavior. The pilot plant is set with a batch of 2.4 kg of mill-balls of 52 mm in diameter. During the tests the pilot-plant was set up to 97 rpm, corresponding to 90 pct of the critical speed, the average relative position of the balls within the mill is illustrated in Figure 2. Several batches containing 2 kg of ore were ground during periods of 24 hours. Samples of homogenized ore of less than 12.7 mm in size, fed the pilot-plant ball-mill. Crushed materials of the ore sampling from the supplier (mine) was analyzed by XRD; thus, silicon oxide (quartz); aluminum-hydro-silicate, smectite-kaolinite (feldspar); calcium oxide (calcite), silver, iron, copper and gold traces were found. The tests were performed according to the following steps: "run-in", the balls were subjected to abrasive wear during 24 h within the pilot-plant ball mill, in fact, in this step actually rush scales and other defects from the surface are removed; grinding, periods of 24 up to 120 h of total grinding time were established. At the end of each grinding period, the mill was unload and re-loaded with fresh ore. Weighing; at the end of each grinding period, the mill was unload and the balls were cleaned and weighed, the weight loss of the balls gave a "benchmark" in order to predict the life of each ball.

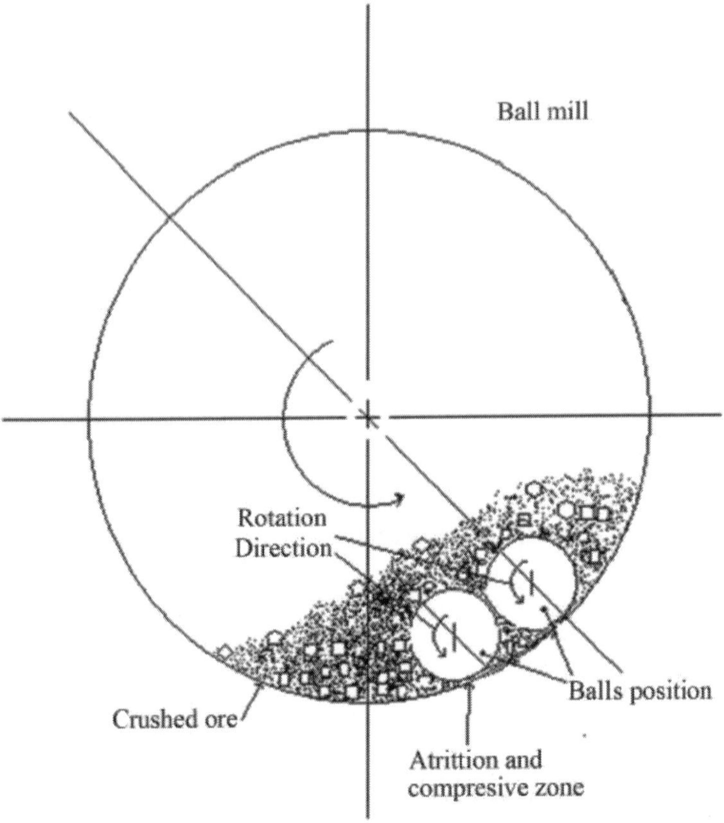

Figure 2. Position the balls within the mill.

3. RESULTS AND DISCUSSION

3.1. Microstructure and Hardness Analysis

Figure 3(a) shows that microstructure of Sample 1 is comprised of dendritic matrix structure along with a homogeneous distribution of eutectic carbides. Detailed observations made on the microstructure of Sample 1 actually revealed characteristic honeycomb ledeburite morphology of eutectic carbides (Figure 3(b)).

Microstructure of Samples 2 to 5 is illustrated in Figures 4(a)-(d). The primary phase in all alloys is comprised of dendritic morphology and eutectic plate-like carbides located at the inter-dendritic spacing (i.e. Figures 4(a)-(d)). According to Figures 4(a)-(d), a finer primary microstructure is revealed in Sample 3 (see Figure 4(b)) in comparison to other samples; in addition, a homogeneous distribution of carbide regions is clearly observed in Sample 3.

According to SEM observations in Figure 5, the microstructure of Sample 1 clearly revealed the morphology of eutectic carbides of ledeburite type (i.e. M_3C, M_7C_3), along with scarce amount of molybdenum carbides (Mo_2C) embedded within a martensite matrix (α'). Further, there was no presence of titanium and vanadium carbides in Sample 1.

In contrast, the microstructure of Sample 3 (Figure 6) is composed of plate-like eutectic carbides with higher fraction of M_7C_3, further accompanied of small carbides of vanadium-titanium (V-Ti)C, and a fraction of molybdenum carbides (Mo_2C). Correspondingly, a fibrous martensitic matrix (α') and small fraction of retained austenite and twinned martensite is revealed.

XRD diffractograms of the studied alloys are observed in Figure 7. According to XRD results, observed phases in both cast alloys (Samples 1 and 3) corresponded to: interdendritic eutectic carbides of M_7C_3-M_3C, Mo_2C carbides, and a primary phase corresponding to martensite (α'). Particularly, Sample 3 was confirmed with presence of (V-Ti)C carbides. Based on XRD results, the formation of carbides is a combination of M_7C_3 surrounded by M_3C type [11] [12] .

Figure 8 shows Rockwell hardness and Vickers microhardness results. Sample 3 resulted in higher values of hardness compared to that obtained in Sample 1. Higher hardness values obtained in Sample 3 are consistent to the refined primary microstructure (Figure 4(b)), the type and fraction of carbides and the presence of twinned martensite.

It is clear that plate-like morphology of M_7C_3 carbides observed in Sample 3, influenced on the elevated hardness; besides, the higher percentage of Cr (i.e. Table 1) actually promoted larger volume fraction of M_7C_3 carbides in Sample 3. Additionally, V and Ti content (Table 1), further helped to refine the martensitic microstructure as well as to increase the measurable hardness owe to formation of (V-Ti)C [10] [22] .

It has been reported that M_7C_3 carbides have shown higher measurable hardness when compared to honeycomb ledeburite M_3C carbides [10] .

3.2. Wear Analysis

Abrasive wear resistance (weight lost) of all samples analyzed in this research is plotted against grinding time, as indicated in Figure 9. According to Figure 9; reduced wear resistance is observed in Sample 1 (owed to weight lost) in the full range of grinding time. In comparison, Samples 2, 4 and 5 resulted in acceptable wear performance. On the other hand, enhanced wear behavior was revealed in Sample 3.

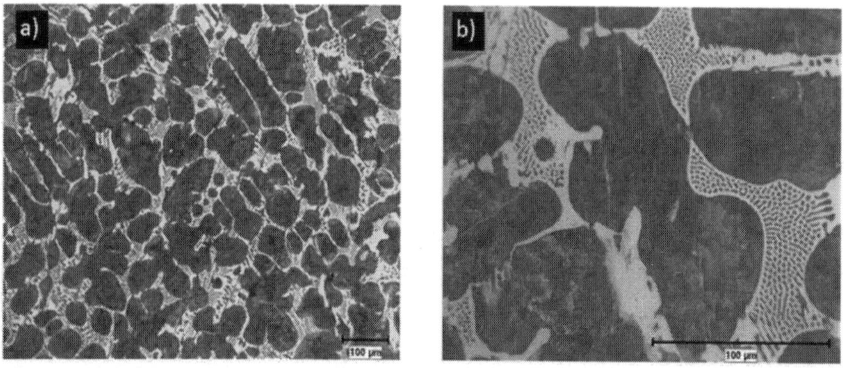

Figure 3. (a) Optical micrographs showing the microstructure of Sample 1; (b) Detailed microstructure depicting a dendritic matrix along with eutectic carbides at the interdendrític spacing.

Based on the above mentioned results (Figure 9); improved abrasive wear resistance is observed in Sample 3. It is clear that the higher fraction of plate-like carbides of the type M_7C_3, the grain refinement (Figure 4(b)), and the elevated hardness values in Figure 8, influenced on the enhanced wear behavior of Sample 3; besides, wear behavior improvement is well supported by the presence of (V-Ti)C carbides. On the other hand, reduced resistance to abrasive wear found in Sample 1; corresponded to presence of honeycomb ledeburite phase within the cast alloy as well as the lowered hardness (Figure 8).

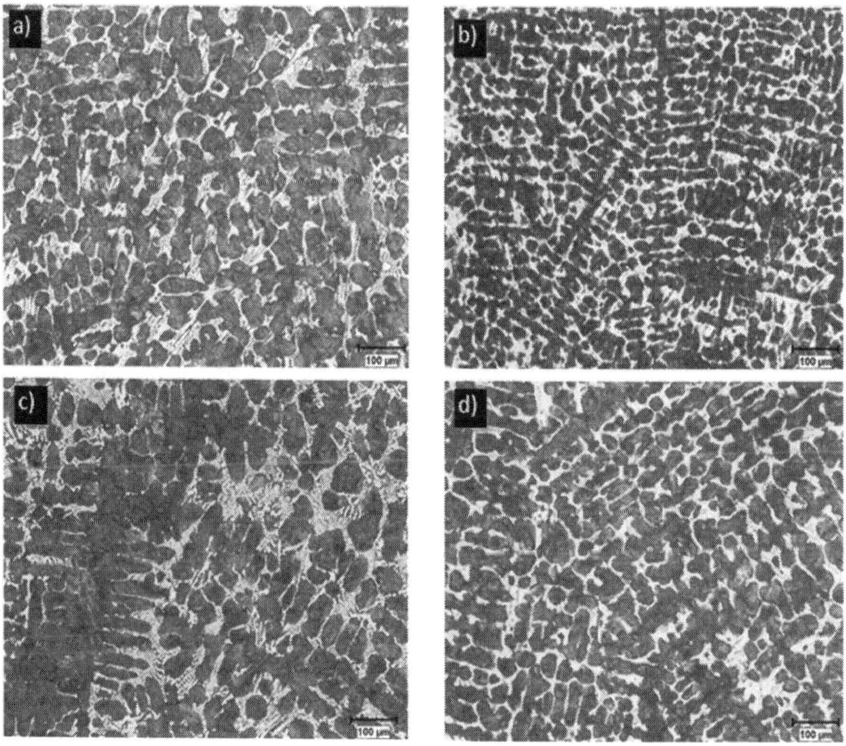

Figure 4. Optical microscope images showing the microstructure of the different cast alloys having a dendritic structure along with eutectic-carbides at the interdendritic spacing for: (a) Sample 2; (b) Sample 3; (c) Sample 4; (d) Sample 5.

Figure 10(a) shows a cross-section obtained along the worn area in Sample 1. Cracking across the bulk car-bide is clearly seen in Figure 10(a) (as indicated by the arrow). Further analysis made through fractography (Figure 10(b)) showed cracking along the subsurface accompanied by a significant amount of delamination owed to maximum stresses generated underneath the surface. In fact, cracks nucleate and propagate under the surface thus fragmenting later. According to the above mentioned, fatigue mechanism is developed and can be stated as the main wear mechanism.

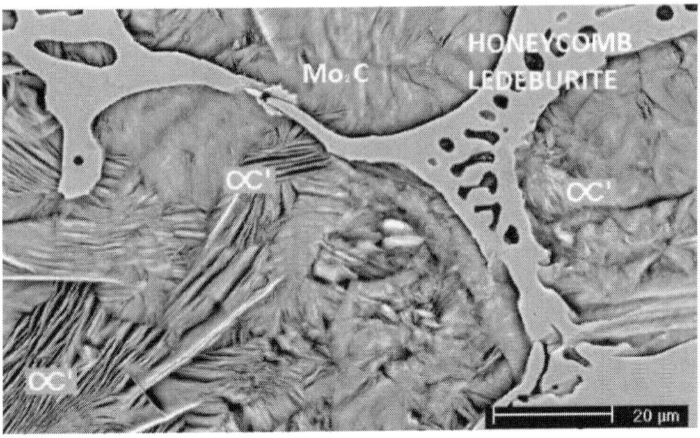

Figure 5. SEM micrographs showing the microstructure of Sample 1: carbides of Mo_2C, along with ledeburitic structure, with M_7C_3 carbides and martensite.

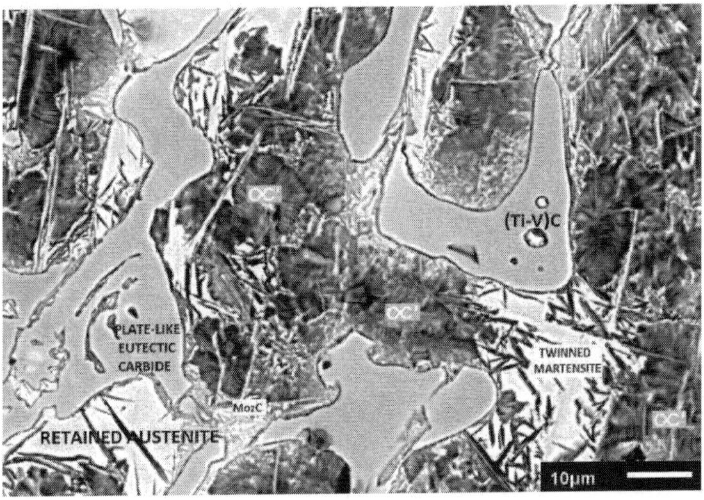

Figure 6. SEM image showing the microstructure of Sample 3: an austenite matrix, fine needles of $(Si, Mo)_2C$, $(V-Ti)C$, M_7C_3 and twinned martensite.

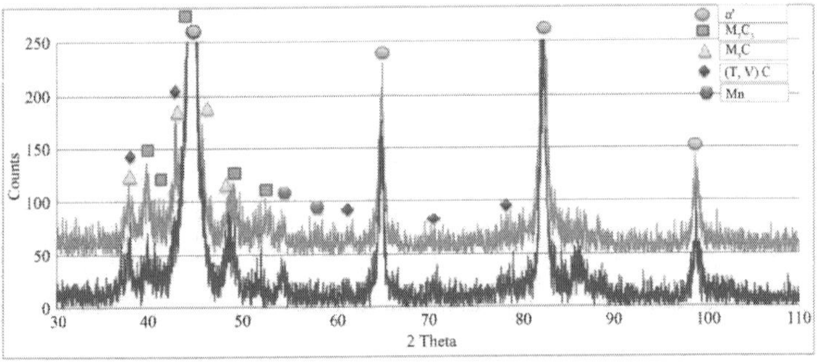

Figure 7. Diffractograms superimposed on Sample 3 (upper green line) to 1 (lower red line).

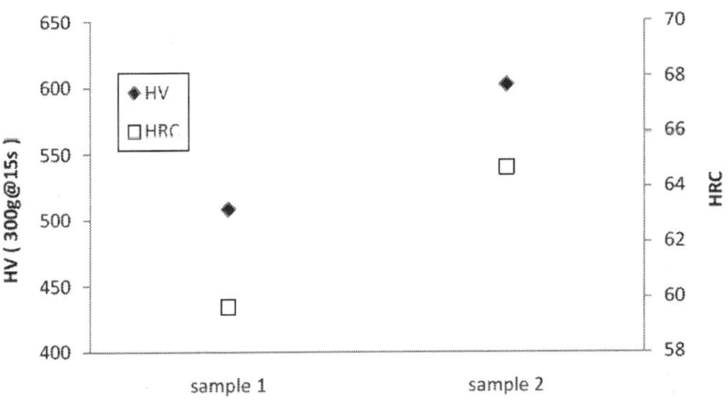

Figure 8. Rockwell hardness (HRC) and microhardness vickers (HV) results for Sample 1 (darker point) and Sample 3 (whiter point).

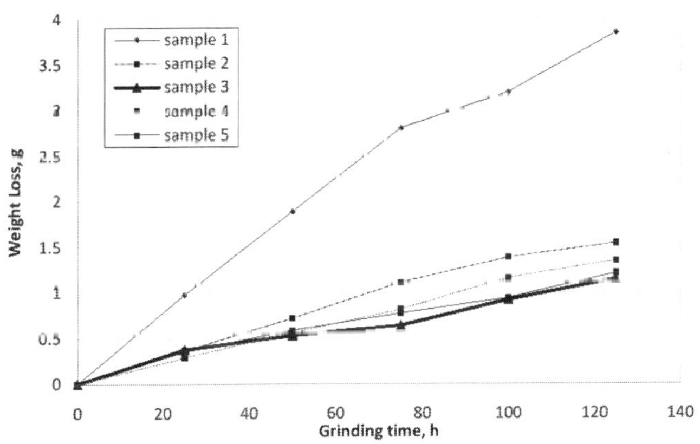

Figure 9. Abrasive wear rate of all samples.

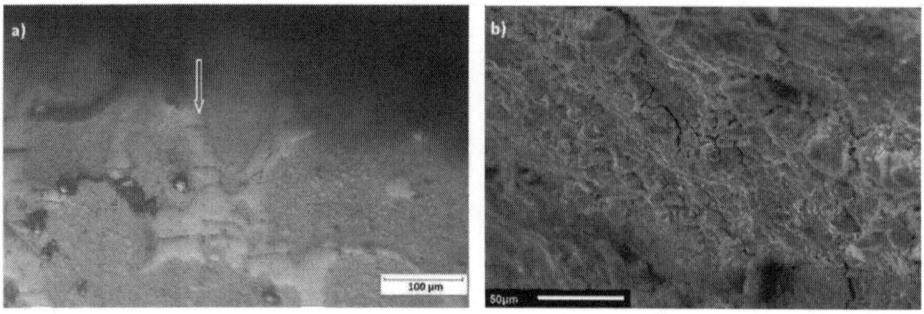

Figure 10. (a) Optical micrograph depicting the cross-section of the wore area in Sample 1; showing carbide cracking; (b) Fractography analysis indicating fatigue failure mechanism along the sub surface.

On the other hand, Figure 11(a) confirmed similar wear mechanism occurring in Sample 3; thus, carbide cracking along with a random location of cracks along the matrix is observed in Figure 11(b). According to the above mentioned, it might be inferred that the predominant wear mechanism in Sample 3 is fatigue. However, delamination was not observed in Sample 3 due to the type of carbides and the refined primary matrix, thus enhancing wear resistance. Carbide cracking in martensitic matrix cast iron balls has been also reported [1] .

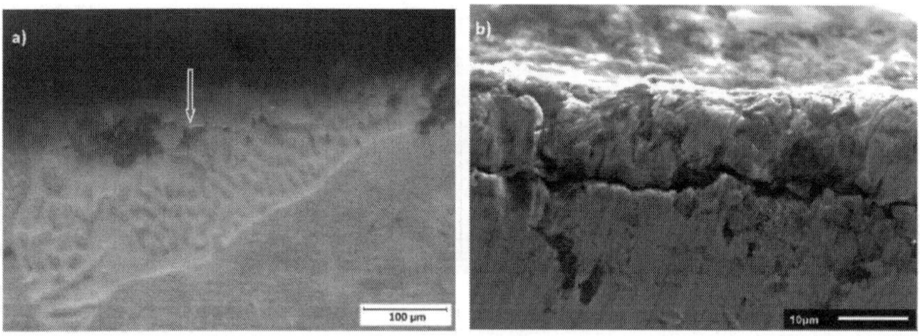

Figure 11. (a) Optical micrograph depicting the cross-section of the wore area in Sample 3 indicting carbide cracking; (b) fractography analysis in Sample 3 indicting fatigue failure mechanism along the subsurface.

Based on the above results, honeycomb ledeburitic microstructure resulted in reduced hardness in Sample 1 (see Figure 8) in comparison to higher hardness in M_7C_3 and M_3C carbides type in Sample 3. It has been reported that chromium carbide (M_7C_3-M_3C) present higher hardness in comparison to cementite (M_3C), thus resulting in good abrasion resistance for those materials [10][23] .

Additions of V and Ti seemed to refine the microstructure in Sample 3 (Figure 4(b)). Vanadium is a strong carbide-forming element if contained in the range 0.1% to 0.5% and is commonly found to form primary and secondary carbides when combined with iron, increasing hardness with V content[10] . Likely, the effect of Ti (in the composition up to 0.38%) is to improve abrasion resistance owe to formation of TiC [22] . Hereby, a combination of massive M_7C_3 and the presence of relatively fraction of (V-Ti)C was highly influencing on the resultant improved wear resistance of low Cr white cast irons.

4. CONCLUSIONS

1) Honecomb ledeburitic (M_3C) microstructure along with the scarce amount of molybdenum carbides (Mo_2C) embedded within a martensite matrix of the studied mill balls resulted in reduced abrasion wear resistance.

2) Improved wear resistance in cast steels mill balls was influenced by the higher fraction of plate-like carbides of the type M_7C_3, the refined martensitic matrix, as well as the elevated measurable hardness.

3) It was found that additions of Ti and V helped on the refinement of the martensitic matrix as well as on the formation of (V-Ti)C carbides, which in turn, influenced on the elevated hardness and the resultant wear behavior.

ACKNOWLEDGEMENTS

The authors acknowledge the assistance of Mr. Ing. Maximino Burgos, grinding balls seller in Mexico, for his contributions to the discussion of the chemical composition. The authors would like to acknowledge FomixConacyt, México; for sponsoring this research.

REFERENCES

1 Albertin, E. and Sinatora, A. (2001) Effect of Carbide Fraction and Matrix Microstructure on the Wear of Cast Iron Balls Tested in a Laboratory Ball Mill. Wear, 250, 492-501.

2 Cikara, D., Rakin, M. and Todic, A. (2009) Cast Steel-SiC Composites as Wear Resistant Materials. FME Transactions, 37, 151-155.

3 Studnicki, A., Kilarski, J., Przybyl, M., Suchoń, J. and Bartocha, D. (2006) Wear Resistance of Chromium Cast Iron—Research and Application. Journal of Achievements in Materials and Manufacturing Engineering, 16, 63-73.

4 Rajagopal, V. and Iwasaki, I. (2006) Corrosion Properties of Cast Iron Ball Materials in Wet Grinding. Corrosion, 48, 124-131.

5 Chenje, T.W., Simbi, D.J. and Navara, E. (2004) Relation between Microstructure, Hardness, Impact Toughness and Wear Performance of Selected Grinding Media for Mineral Ore Milling Operations. Materials & Design, 25, 11-18.

6 Sampathkumaran, P., Seetharamu, S., Ranganathaiah, C., Raj, J.M., Pujari, P.K., Maheshwari, P., Dutta, D. and Kishore (2011) Surface and Bulk Defects in Cr-Mn Iron Alloy Cast in Metal and Sand Moulds: Characterization by Positron Annihilation Techniques. Journal of Surface Engineered Materials and Advanced Technology, 1, 136-143.

7 Atabaki, M.M., Jafari, S. and Abdollah-Pour, H. (2012) Abrasive Wear Behavior of High Chromium Cast Iron and Hadfield Steel—A Comparison. Journal of Iron and Steels Research, 19, 43-50

8 Weber, K., Regener, D., Mehner, H. and Menzel, M. (2001) Characterization of the Microstructure of High-Chromium Cast Irons

Using Mösbauer Spectroscopy. Materials Characterization, 46, 399-406.

9 Thorpe, W.R. and Chicco, B. (1993) The Fe-Rich Corner of the Metastable C-Cr-Fe Liquidus Surface. Metallurgical Transactions A, 24, 981-988.

10 Breyer, J.P. and Walmag, G. (2002) Metallurgy of High Chromium-Molybdenum White Iron and Steels Rolls. Rolls for the Metalworking Industries—ISS, Warrendale, 29-40.

11 Schon, C.G. and Sinatora, A. (1998) Simulation of Solidification Paths in High Chromium White Cast Irons for Wear Applications. Calphad, 22, 437-448.

12 Hugget, P. and Ben-Nissan, B. (2007) Development of a Low Melting Point White Cast Iron for Use in Composite Alloy Manufacture. Materials Forum, 31, 16-23.

13 Agunsoye, J.O., Isaac, T.S. and Abiona, A.A. (2013) On the Comparison of Microstructure Characteristics and Mechanical Properties of High Chromium White Iron with the Hadfield Austenitic Manganese Steel. Journal of Minerals and Materials Characterization and Engineering, 1, 24-28.

14 Agunsoye, J.O., Aigbodion, V.S. and Sanni, O.S. (2011) Effect of Heat Treatment on Microstructure and Mechanical Properties of NF6357A Cast Alloy for Wear Resistance Application. Journal of Minerals & Materials Characterization & Engineering, 10, 1077-1086.

15 Albertin, E.B. and Sinatora, A. (2001) Effect of Carbide Fraction and Matrix Microstructure on the Wear of Cast Iron Balls Tested in a Laboratory Ball Mill. Wear, 250, 492-502.

16 Dogan, O.N., Laird II, G. and Hawk, J.A. (1995) Abrasion Resistance of the Columnar Zone in High Cr White Cast Irons. Wear, 181-183, 342-349.

17 Dogan, O.N. (1996) Columnar to Equiaxed Transition in High Cr White Iron Castings. Scripta Materialia, 35, 163- 168.

18 Maldonado-Ruiz, S.I., Lopez, D., Velasco, A. and Colas, R. (2004) Microstructural Evaluation of Wear Resistant High Chromium, High Carbón Cast Irons. Materials Science and Technology, 20, 393-398.

19 Maldonado-Ruíz, S.I., Martínez, D.I., Velasco, A. and Colás, R. (2005) Wear of White Cast Irons by Impact of Direct Reduced Iron Pellets. Wear, 259, 361-366.

20 (2010) Design-Ease 7.1, Stat-Ease, Inc™.

21 Albertin, E. and De Moraes, S.L. (2007) Maximizing Wear Resistance of Balls for Grinding of Coal. Wear, 263, 43-47.

22 Arikan, M.M., Cimenoglu, H. and Kayali, E.S. (2001) The Effect of Titanium on the Abrasion Resistance of 15Cr-3Mo White Cast Iron. Wear, 247, 231-235.

23 Werquin, J.C. and Caillard, J.C. (1988) New Vertically Spum Cast Rolls for Hot and Cold Rolling and Their Ability to Satisfy the New Demands of Current Rolling Practices. 30th MWSP Conference Proceedings, 26, 311-336.

CHAPTER 9

A Review on the Properties of Iron Aluminide Intermetallics

Mohammad Zamanzade [1,*], Afrooz Barnoush [2] and Christian Motz [1]

[1] Department of Material Science and Methods (MWW), Saarland University, Saarbrücken D-66041, Germany

[2] Department of Engineering Design and Materials, Norwegian University of Science and Technology, Trondheim No-7491, Norway

ABSTRACT

Iron aluminides have been among the most studied intermetallics since the 1930s, when their excellent oxidation resistance was first noticed. Their low cost of production, low density, high strength-to-weight ratios, good wear resistance, ease of fabrication and resistance to high temperature oxidation and sulfurization make them very attractive as a substitute for routine stainless steel in industrial applications. Furthermore, iron aluminides allow for the conservation of less accessible and expensive elements such as nickel and molybdenum. These advantages have led to the consideration of many applications, such as brake disks for windmills and trucks, filtration systems in refineries and fossil power plants, transfer rolls for hot-rolled steel strips, and ethylene crackers and air deflectors for burning high-sulfur coal. A wide application for iron aluminides in industry strictly depends on the fundamental understanding of the influence of (i) alloy composition; (ii) microstructure; and (iii) number (type) of defects on the thermo-mechanical properties. Additionally, environmental degradation of the alloys, consisting of hydrogen embrittlement, anodic or cathodic dissolution, localized corrosion and oxidation resistance, in different environments should be well known. Recently, some progress in the development of new micro- and nano-mechanical testing methods in addition to the fabrication techniques of micro- and nano-scaled samples has enabled scientists to resolve more clearly the effects of alloying elements, environmental items and crystal structure on the deformation behavior of alloys. In this paper, we will review the extensive work which has been done during the last decades to address each of the points mentioned above.

KEYWORDS

aluminide intermetallics; alloy design; brittleness and ductility; corrosion; oxidation; hydrogen embrittlement

1. INTRODUCTION

Transition metal (TM)—aluminide intermetallics including TiAl, NiAl, FeAl and Fe_3Al have unique properties, e.g., high melting points, enhanced oxidation resistance, relatively low density, and can be used as soft magnetic materials [1,2,3,4,5,6,7]. Early TM-aluminides have fcc-based crystal structure in contrast to the bcc-based crystal structure of late transition metal alloys. Due to the strongly attractive chemical bonding between the bi-metallic species, they are ordered and have stoichiometry. However, the energy of interatomic bonds differs from the early TM (TiAl, VAl) to the late TM alloys (CoAl, NiAl and FeAl). The calculated heats of formation for aluminides with equiatomic composition are plotted inFigure 1 [8]. The heat of formation is remarkable for the case of Ni aluminides in comparison to the Ti aluminides [9]. However, in both cases Al atoms act as an electron donor. The middle TM aluminides are shown to have much less charge transfer and the lowest degrees of ordering. In the case of late TM alloys, charge transfer from Al to the TM and hybridization between Al sp and transition metal d-states are the driving force in bonding [8]. Therefore, the Fe-Al neighbors are energetically favored and Al-Al neighbors are avoided.

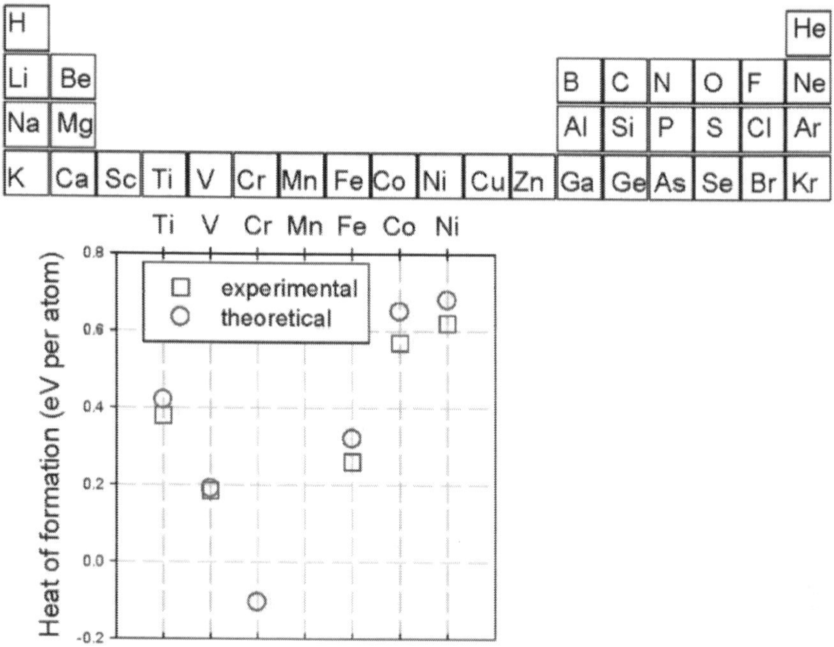

Figure 1. Heats of formation for transition-metal aluminides (with equiatomic composition). Square and circle symbols are experimental and theoretical values, respectively. Re-plotted from the reference [8].

Aluminides were first aimed to be used at high temperatures instead of Ni-based superalloys. Hence, some of the thermo-mechanical properties of the aluminides have been reviewed and compared with the CMSX-4 Ni-based superalloys inFigure 2 [10,11,12,13]. The six major advantages of aluminides for high temperature structural applications over the Ni-based superalloys include: (i) higher melting temperatures where the melting temperatures define the upper limit of the service temperature and are indicators of the temperature range at which diffusion-controlled processes start to dominate; (ii) lower densities where the lower densities (especially in the case of γ-TiAl intermetallics) result in lower operating stresses that make it possible to fabricate smaller and lighter components which, in turn, result in better engine accelerations due to the lower mass of the rotating parts; (iii) better oxidation resistance because of the high aluminum content; (iv) lower ductile to brittle transition temperature (DBTT); (v) similar thermal expansion coefficients as the bcc steels and (vi) low costs of production, since they do not generally incorporate rare and strategic elements [10,11,12,13]. In contrast, the three negative features of the intermetallics at elevated temperatures are (i) low strength; (ii) limited creep resistance; and (iii) high thermal conductivity. At low to moderate temperatures, most of the intermetallics additionally suffer from poor ductility (Figure 2) and low fracture toughness [14,15]. These problems significantly impede the wide usage of intermetallics because the machining of alloys (at low temperature) becomes very difficult.

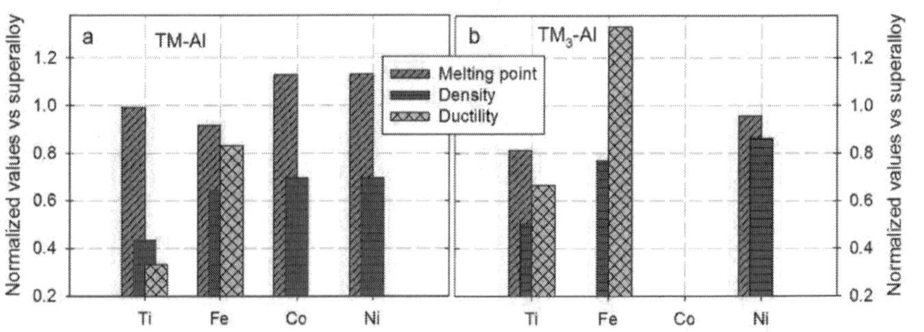

Figure 2. Mechanical properties of transition-metal aluminides [10,11,12,13]. (**a**) Equiatomic composition (e.g., 1 Fe:1 Al) and (**b**) three TM: 1 Al compositions. Melting temperature of CMSX-4 Ni-based super-alloy ≈ 1450 °C, Density ≈ 8.70 gr/cm³, Ductility ≈ 3%.

Liu *et al.* [16] presented the ductile behavior of aluminides in dry oxygen, where the fracture strain is about 17.6% for Fe-36.5 at. % Al. In general, increasing the Al concentration decreases the density of materials and enhances the protective oxide layer at high temperatures

[17,18]. However, the existence of high aluminum concentration has negative side effects [19,20]. The reaction of Al atoms with water results in the production of hydrogen atoms, which are responsible for the low ductility of Fe-Al based intermetallic alloys in moisture-containing atmospheres [21,22,23,24] (Figure 3).

The rate of a charge-transfer process such as that of hydrogen adsorption depends on the potential across the sample-solution (environment) interface, which can be electrochemically varied in a controlled manner and also by the engineering of the interface. A passive oxide film, for example, causes a drop in the electric current flow. Formation of protective oxide layers on surfaces by pre-oxidizing the iron aluminides may influence the hydrogen adsorption and dissolution in the material and enhance ductility of samples. Furthermore, the thicknesses, crystal structure (if it is amorphous or crystalline), electronic or ionic conductivity, type (if it is n-type or p-type) and numbers of donors in the passive layer may vary based on the alloying elements.

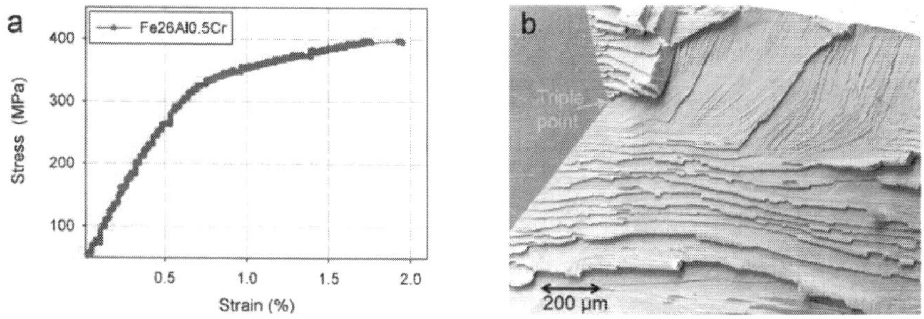

Figure 3. (**a**) Shows low ductility of Fe-Al aluminides; (**b**) Cleavage-like fracture surface of a Fe-26Al-0.5Cr intermetallic after doing tensile test in air. Data from the reference [23].

McKamey *et al.* [25] observed an approximate 8%–10% increase in ductility at room temperature with the addition of 6 at. % chromium. The reasons for the increased ductility after Cr addition are thought to be caused by: (i) the influence of Cr on the bulk properties of binary alloys, such as facilitating the dislocation cross-slipping, solid solution softening and an increment in cleavage strength, and/or (ii) the effect of Cr on the surface properties through the contribution of chromium oxide into the passive layers and the decrement of the kinetics of water reduction reactions, which leads to a reduction of hydrogen formation/adsorption [26,27,28]. However, the intrinsic complexities of the macro-scale experimental examination make it impossible for us to precisely explore the mechanisms of hydrogen embrittlement in the iron aluminides [29]. In contrast, with the aid of recent local *in-situ* micro-, nano-mechanical testing methods, we can reduce some sources of scatters such as material in-homogeneities during the measurements. Here we will discuss the abilities of *in-situ* nanoindentation measurements [24,29] and *in-situ* compression tests of micro-pillars [22] for evaluating the effects of hydrogen on various mechanical properties such as elastic modulus, dislocation nucleation and mobility. In this review paper, however, it is impossible to develop any of these topics in great detail. Nevertheless, we hope that each section contains enough information for the average reader to understand what has been achieved in that field, as well as satisfactory references to provide further reading if necessary.

2. CHARACTERISTICS OF FE-AL INTERMETALLICS

2.1. Phase Diagram

The phase diagrams for binary iron aluminides were studied using different techniques, e.g., dilatometric measurement [30], electrical resistivity and specific heat measurement [31], measurement of elastic modulus and magnetic properties [32,33], X-ray diffraction methods [34]; and, more recently, the phase diagram was intensively studied with TEM [35]. The first well-accepted binary Fe-Al phase diagram was reported in 1982 [36]. The phase boundaries and transitional temperatures may differ slightly from one reference to the other. The reason for this uncertainty is the anomaly in the physical properties such as Young's modulus, thermal expansion and electrical resistance. Some possible explanations for the observed irregularities including other long-range ordered states, short-range ordering, two-phase structure, material impurities such as carbide precipitation and quenched-in vacancies are proposed [35,37,38,39]. The different phases and their lattice structure and lattice parameters are listed in Table 1 [9,40,41,42,43,44,45,46].

Table 1. Different Fe-Al phases.

Phase	Label in Figure 4	Pearson Symbol	Space Group	Prototype	Lattice Parameters (nm)	Elastic Constants (eV/Å³)
Liquid	*L*					
Al	*Al*	*cF4*	$Fm\bar{3}m$	*Cu*	$a_0 = b_0 = c_0 =$ 0.40496 [43]	C_{11} = 0.6492, C_{12} = 0.4619 C_{44} = 0.2684 [44]
fcc (*γFe*)	*γFe*	*cF4*	$Fm\bar{3}m$	*Cu*	$a_0 = b_0 = c_0 =$ 0.36599 [40]	–
bcc (*αFe*)	*A2* (*pm*)* *A2* (*fm*)**	*cI2*	$Im\bar{3}m$	*W*	$a_0 = b_0 = c_0 =$ 0.28665 [43]	C_{11} = 1.4357, C_{12} = 0.8426 C_{44} = 0.73 [44]
Fe_3Al	$D0_3$	*cF16*	$Fm\bar{3}m$	BiF_3	a0 = b0 = c0 = 0.2895 [40], 0.5904/2 [44], 0.5792/2 [43]	[44]: C_{11} = 0.945, C_{12} = 0.892, C_{44} = 0.788 [45]: 1.067, 0.822, 0.815
FeAl	*B2* (*pm*)	*cP8*	$Pm\bar{3}m$	*CsCl*	$a_0 = b_0 = c_0 =$ 0.291 [41,43,46], 0.283 [9], 0.3031 [44]	[46]: C_{11} = 1.2, C_{12} = 0.75, C_{44} = 0.73 [9]: 1.8, 0.8, 1.029 [44]: 0.883, 0.846, 0.691
Fe_5Al_8***	ε ($D8_2$)	*cI52*	$I\bar{4}3m$	Cu_5Zn_8	n.a. [42]	–
$FeAl_2$	$FeAl_2$	*aP18*	*P*1	$FeAl_2$	a_0 = 0.4872, b_0 = 0.6459, c_0 = 0.8794, α = 91.76 β = 73.35, γ = 96.89 [41]	–
Fe_2Al_5	Fe_2Al_5	*oC?*	*Cmcm*	–	a_0 = 0.7652, b_0 = 0.6463, c_0 = 0.4229 [41]	–
Fe_4Al_{13}	Fe_4Al_{13}	*mC102*	*C2/m*	–	n.a. [41]	–

* Paramagnetic; ** Ferromagnetic; *** Al-rich intermetallics such as $FeAl_2$, Fe_2Al_5 and Fe_4Al_{13} have lower densities in comparison to FeAl and Fe_3Al but are not good candidates for structural applications because of their small homogeneity ranges, complex crystal structures and brittle behavior. Therefore, we will not focus on them in this text.

Figure 4 [47] shows the extension of the solid solution of Al in Fe from 0 up to 45 at. % Al at high temperatures. At low temperatures, this region is subdivided into three parts. The first part consists of disordered alloys up to 18.75 at. % Al at room temperature. At about 25 at. % Al and temperatures below 545 °C, the intermetallic phase Fe_3Al is formed. The next upcoming phase at higher Al contents is FeAl. Both Fe_3Al ($D0_3$) and FeAl (B2) phases are ordered forms on the body-centered cubic (bcc) lattice and are separated from the disordered A2 phase by first- or second-order transitions.

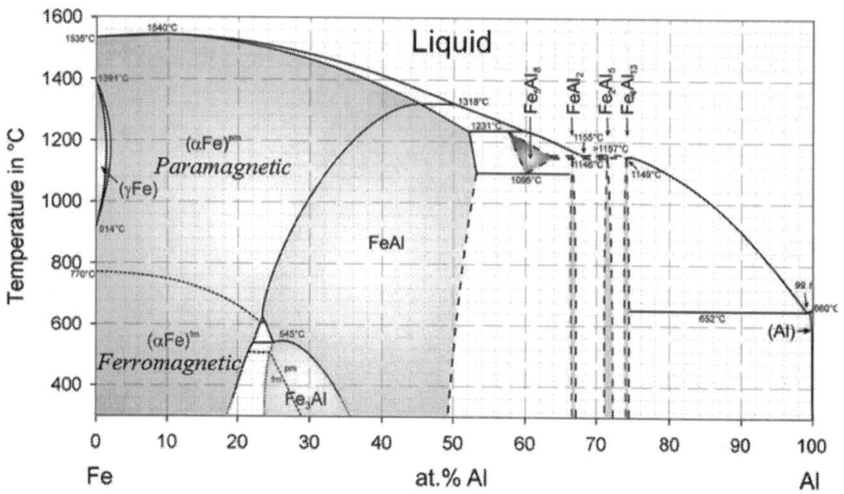

Figure 4. Fe-Al phase diagram. In the phase diagram some of the solubility lines are plotted with dashed lines because they are still not well determined. Data from reference [47].

The X-ray diffraction method was used for studying the distribution of iron and aluminum atoms in the crystal lattice of FeAl (B2) and Fe_3Al ($D0_3$) phases [33,48,49]. The lattice sites in a unit cell are classified into three kinds, α, β and γ as shown in Figure 5a. The probability for each one of these lattice sites to be occupied by an Al atom at room temperature has been plotted against the Al composition, as seen in Figure 5b [48,49]. In the FeAl (B2) crystal lattice, in the stoichiometric composition, Al atoms occupy the body center sublattices (β and γ) and the Fe atoms occupy the corner sites (α sublattice). In the Fe_3Al ($D0_3$) structures, however, the Fe atoms occupy both α and β sublattices, while the Al atoms occupy the γ sublattice. In a $D0_3$ structure, each Fe atom on the α sublattice is surrounded by four Fe and four Al atoms as nearest neighbors (NN), but the Fe atoms on the β sublattice and the Al atoms on the γ sublattice have eight Fe atoms as NNs. This difference significantly influences the defect formation energies [50].

2.2. Point Defects in the Super Cells

The B2 structure has the ideal stoichiometry of 50 at. % Fe-50 at. % Al and the $D0_3$ super structure has the ideal stoichiometry of 25 at. % Fe-75 at. % Al. However, the formation of lattice defects, e.g., vacancies and/or anti-sites, allows for the large range of deviations from the stoichiometric composition. The type and concentration of the point defects are important issues in Fe-Al intermetallics because the constitutional vacancies, or anti-sites, influence the thermo-mechanical, magnetic and electrical properties of the intermetallics. Moreover, the point defects control diffusion-assisted processes, such as creep. The effect of quenched-in thermal vacancies on solid solution strengthening, yield stress and hardness of alloys was studied in detail in the literature we refer to [5,51,52,53,54]. It has been shown that vacancies could act as obstacles to dislocation movement. Additionally, strong interactions between structural defect and hydrogen could significantly influence the hydrogen concentration and penetration rates in iron aluminides and cause severe mechanical degradation. Furthermore, the knowledge of point defect energetics is essential for the proper estimation of the formation (excess) energy of extended one-dimensional (1D) or two-dimensional (2D) defects, such as dislocations or grain boundaries, with local off-stoichiometry.

Point defect structure in intermetallics has been studied extensively with the help of various theoretical methods such as nearest-neighbor pair-wise interactions [55], embedded atom potentials [56] and local-density-functional (LDF) models [57]. The estimated formation energies of different defects in B2 and $D0_3$ super-lattices are presented in Table 2. In a B2 super-lattice, the low formation energies of anti-sites at the Fe and Al sites ([FeAl] and [AlFe]) and also Fe mono-vacancy ([VFe]) cause high concentrations of these forms of defects. In fact, [VFe] and [AlFe]are dominant defect types for Al-rich Fe-Al. For Fe-rich Fe-Al, on the other hand, the constitutional [FeAl] sites are the main defect type. However, the presence of [VAl] is thermodynamically not favored due to its very high energy of formation.

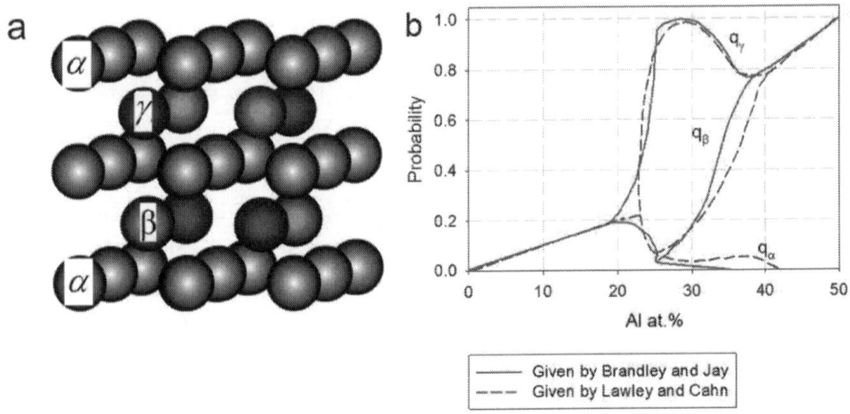

Figure 5. (**a**) Atomic arrangement in a B2 or $D0_3$ super-lattice; (**b**) Occupation probabilities of the lattice sites by Al in a Fe-Al system [48,49].

As it is shown in Table 2, six various forms of point defects could exist on the $D0_3$ sublattices; Fe atoms on the γ sublattice (anti-site Fe atoms), Al atoms on the *α or β* sublattices (anti-site Al atoms) and vacancies on the α, β or γ sublattices. In contrast to the B2 structures, the effective formation energies of all different local defects including vacancies and anti-site atoms vary only slightly with composition in the $D0_3$ structure. However, due to the lower vacancy formation energies in the [VFe−α] in comparison with [VFe−β], the probability of the [VFe−α] is higher. Additionally, the calculated formation energy of [VAl] is not very high; therefore, formation of [VAl] along with [VFe−α] is possible.

Along with studying the thermodynamically stable types of defects, the kinetics of defect formation or annihilation are also essential. The A2 crystal structure has a much larger vacancy formation enthalpy, which causes a much lower vacancy concentration compared to the ordered structures (Figure 6a) [39]. The lowest enthalpy and, therefore, the highest concentration of thermal vacancies can be found for the B2 crystal structure [39]. The $D0_3$ structure will have a lower thermal vacancy concentration (in comparison with the B2 structure); it is what we would expect based on the DFE calculations. Most of calculations did not consider the interaction of defects. In reality, however, the vacancy concentrations of both B2 and D03 superlattices are well above thc prcdicted values based on the theoretical calculations. This is due to the interaction of different vacancies and sublattices, which form defect clusters. The higher effective formation volume of the defects in the B2 phases (1.4 Ω), in comparison with one atomic volume Ω, was observed in reference [58]. In the ordered B2 phase, the main defect types are changing from triple defects at low temperature to double vacancies at higher temperature and Al content [58]. In aluminides with less than 35 at. % Al, the type of defects at low temperatures are typically mono-vacancies. Figure 6b [59,60,61,62] shows the changes of the equilibrium vacancy concentration (Cv) of samples with different Al contents at different temperatures based on the experimental approaches. Obviously, the increase of either temperature or Al content will increase Cv.

Table 2. The effective formation energies of all different local defects including vacancies and anti-site atoms in eV.

Defects	$D0_3$	B2
$[V_{Fe\text{-}\alpha}]$	1.25 [63], 1.583 [44]	0.97 [57], 1.06, [64], 0.80 [65], 0.653 [66]
$[V_{Fe\text{-}\beta}]$	2.27 [63], 1.388 [44]	
$[V_{Al}]$	1.4 [63], 2.221 [44]	4.00 [57], 3.46, [64] 2.80 [65], 1.493 [66]
$[Fe_{Al}]$	0.430 [44]	1.03 [57], 0.78[65], 0.95 [66]
$[Al_{Fe\text{-}\alpha}]$	0.047 [44]	0.95 [57], 0.76 [65], 1.03 [66]
$[Al_{Fe\text{-}\beta}]$	0.218 [44]	

The Cv in the materials can be evaluated as follows:

$$C_v = \exp\left[-\frac{\left(E_v^f + pV_v^f - TS_v^f\right)}{k_B T}\right] \quad (1)$$

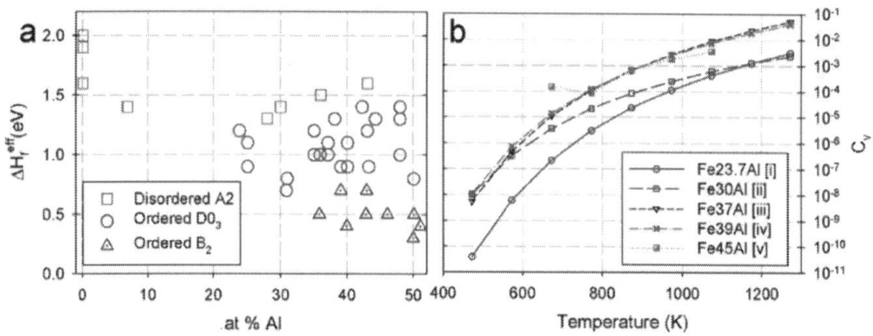

Figure 6. (**a**) Vacancy formation enthalpy for Fe-Al alloys as a function of Al concentration and according to the state of order of the materials [39]; (**b**) Vacancy concentration of Fe-Al alloys based on the experimental approaches (i) [59]; (ii) [60]; (iii) [61]; (iv) [61]; and (v) [62].

Where p is the pressure, E_v^f, V_v^f and S_v^f are the energy, volume and entropy of formation of a vacancy, respectively. The migration enthalpy of the defects enhances with increasing Al content; at a constant temperature, the migration of vacancies seems to be very slow, particularly for the high-Al-content alloys. This is because in ordered alloys, unlike in pure metals in which self-diffusion occurs by random vacancy motion, self-diffusion is not possible since it would disrupt the ordering. The possible diffusion mechanisms in the B2 structures are single vacancy transport, triple defects mechanism and the six-jump cycle. The six-jump vacancy model lets diffusion occur exclusively by nearest neighbor vacancy jumps, though diffusion occurs mostly via nearest neighbor jumps into vacant sites. However, both high migration enthalpy and low formation energy of vacancies dictate the existence of large concentrations of thermal vacancies at high temperatures and the quenching of these thermal vacancies is easy.

2.3. Dislocations in Fe-Al

Two slip directions were observed at low temperatures in a B2 super-lattice while the slip plane remains {110}. The slip direction for NiAl and CoAl is <100> while the slip direction for FeAl and Fe_3Al is <111>. The difference between the active slip directions affects the ductility of intermetallics significantly. NiAl and CoAl only have three independent slip systems, which is less than the minimum number of slip systems needed for the plastic flow of material in the polycrystalline materials based on the Von Mises criterion. Therefore, no ductility at room temperature is expected for these alloys. Since the number of slip systems in the Fe-Al intermetallics is more than the required five independent slip systems, iron aluminides are intrinsically ductile. The burgers vector of the B2 iron aluminides is two times longer than that of a normal dislocation in the bcc structure. The super-lattice dislocation splits into two super-partials, each with a $a/2\langle 111\rangle$ burgers vector (b), separated with an anti-phase boundary (APB) [67]. In Fe_3Al with $D0_3$ structure, a super lattice dislocation with a burgers vector of $\langle 111\rangle$ is known to be dissociated into four super-partial dislocations with $b=a/4\langle 111\rangle$, bound by two types of anti-phase boundaries (APBs): the nearest-neighbor APB (NNAPB) and the next-nearest neighbor APB (NNNAPB) [68]. As it is shown in Figure 7, uncoupled and paired super-partials glide trailing the NNAPB and NNNAPB, respectively. After initiation of the fourth super partial, no APB will be left behind. The surface tension of the NNAPB after initiation of a $a/4\langle 111\rangle$ may pull back the super-partials during unloading and hence cause pseudo-elasticity. Yasudo *et al.* [69] found a significant shape memory effect in $D0_3$ alloys with 23 at. % Al for a wide range of temperatures. At this concentration of Al, the super-partials independently glided. The recovery ratio depends on the Al content and temperature. At 24.7 at. % Al, super-partials moved individually, dragging the NNAPB at room temperature, while at 473 K four-fold dissociated super-dislocations were observed [67,69]. With an Al concentration above 25 at. %, the APB energies increase as long as a homogeneous $D0_3$ structure is present in the Fe-Al phase diagram [70]. At Fe-28.0 at. % Al, a paired super-partial $a/2\langle 111\rangle$glide and a NNNAPB is left behind [68,71]. Table 3 [72,73,74] summarizes the energy of NNAPB (γNNAPB) and NNNAPB (γNNNAPB) of alloys with different Al content.

The preferred slip plane for both B2 and $D0_3$ is {110}. The frequent short distance double cross-slip processes onto {112}-planes and back onto {110}-planes were observed as well [75,76]. It produces wide slip bands in localized areas at temperatures below 350 K, while adjacent areas remain free of mobile dislocations. The prominent screw character of dislocations was seen, as well as a great number of dipoles in the deformed areas, which resulted from frequent cross-slip events.

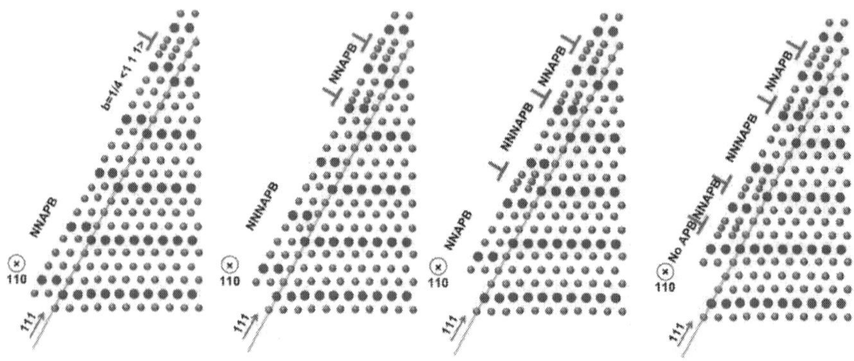

Figure 7. Schematic drawing of the initiation of partial dislocation and formation of anti-phase boundaries in a $D0_3$ super-lattice. Al and Fe atoms are presented in red (bigger) and blue spheres, respectively.

Table 3. The APB energies of $D0_3$ intermetallics at room temperature [72,73,74].

Alloy (at.%)	γ_{NNAPB}	γ_{NNNAPB}	Ref.
Fe-26.3Al	77 ± 12	85 ± 16	[72]
Fe-26.9Al	68 ± 12	72 ± 13	[73]
Fe-27.8Al	76 ± 11	53 ± 11	[73]
Fe-28Al	73 ± 7	80 ± 7	[74]
Fe-28.3Al	79 ± 10	64 ± 10	[73]

2.4. Mechanical Properties at High Temperatures

The effect of Al content on the yield stress of binary alloys at three different temperatures is presented in Figure 8a [37]. At low to moderate temperatures (T = 773 K), an outstanding enhancement of the yield stress was observed for the $D0_3$super structure with stoichiometric concentration. Another remarkable mechanical property of iron aluminides is their anomalous increase in yield strength with increasing temperature up to ≈900 K (Figure 8b [77]), which has been observed in single- [78] and polycrystals [79]. There has been much debate about the origin of the observed stress anomaly in the flow stress. It appears to be controlled by several different mechanisms. These mechanisms include anti-phase boundary relaxation leading to dislocation drag, cross-slip of screw dislocations, changing of slip directions from <111> to <001>, super-dislocation climb locking and vacancy hardening [38,39,76,78,79,80]. Moreover, the Young's modulus of the binary alloys has been studied in detail in a wide range of temperatures (Figure 9) [80,81]. It has been shown that the Young's modulus of Fe-Al intermetallics with $D0_3$ super structure is minimum at room temperature. The Young's modulus increases with the enhancement of the Al content in each super lattice. It shows the effect of aluminum on the strengthening of the interatomic bonds. Interestingly, the Young's modulus of aluminides with high Al content decreases significantly with the increase of temperature (about 40%), while this decrease is less in the aluminides with less Al content (about 25%).

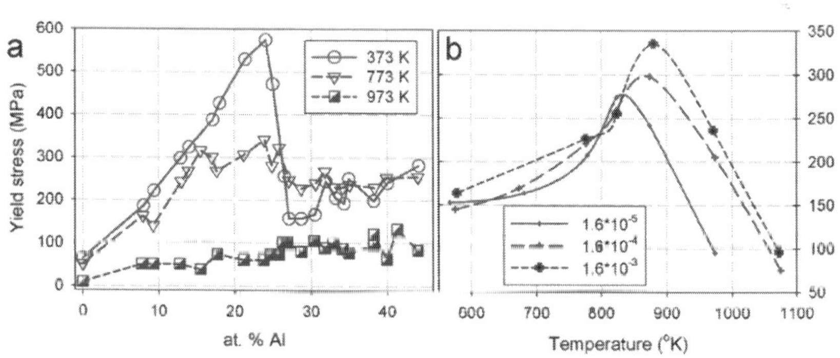

Figure 8. (**a**) Shows the influence of temperature and Al content on the yield stress of binary intermetallics [37]; (**b**) Shows stress anomaly and strain rate sensitivity of Fe-39.5 at. % Al intermetallics for a single slip system–oriented sample [77].

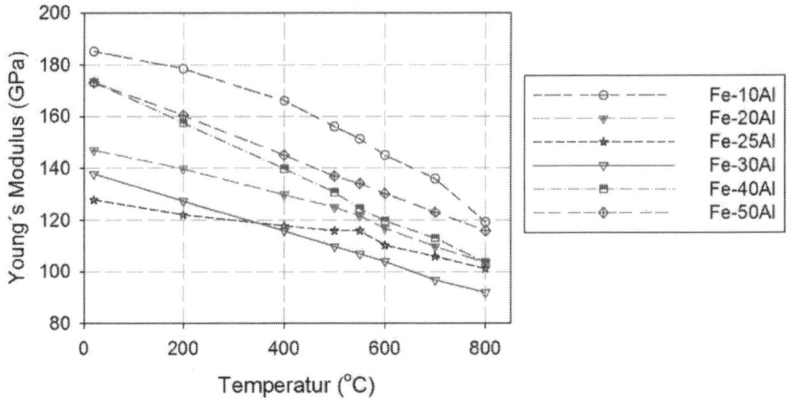

Figure 9. The effect of Al content on the Young's modulus at different temperatures [79,81].

2.5. Alloying Elements

To improve the mechanical, thermal or electrochemical properties of binary Fe-Al intermetallics, one can add some different selected alloying elements to the binary alloys. Some metallic elements such as Si, V, Cr, Mn, Co, Ni, Cu and Zn have large solid solubility [37,82,83] and others such as Zr, Nb and Ta have restricted solid solubility in Fe-Al intermetallics [83,84]. It has been shown that the β sublattice sites are preferred by Ti, W, V, Cr, Mo and Zr whereas Si atoms prefer the γ sublattice in a Fe3Al super-structure [40,85]. The effect of different ternary alloying elements on the formation energies (Ef), relative changes of lattice parameters and Young's modulus were determined using experimental approaches as well as computer simulations (see Table 4) [40,86,87,88].

The DO3→B2 phase transformation which happens at temperatures of about 550 °C may cause detrimental effects on the mechanical properties [35]. To increase the transition temperature, we can add different ternary alloys such as Nb, Ti, Mo and V. Cr has a very slight effect on temperature transition while V, Mo, Ti and Nb enhance the transitional temperature for 34, 36, 57 and 62 K/at. %, respectively. Furthermore, Stein *et al.* [35] evaluated the effect of two alloying elements on the transition temperature. They found that the two elements may have additive influence at least in the range of studied compositions.

The effects of different alloying elements, namely Cr, Mo, Ti and V, on the solid solution strengthening at various temperatures are summarized in Figure 10 [37]. Increase of the yield stress relates to the temperature, type and amount of the solute atoms. While Cr does not significantly influence the yield stress at high temperatures, V, Mo and Ti enhance the flow stress about 500%–600% in comparison to the binary alloy at 800 °C.

Table 4. Effect of ternary alloying elements on the properties of Fe_3Al intermetallics.

–	Formation Energy at Different Sublattices (meV/atom) [40]			Relative Changes of the Lattice Parameter with Respect to the Solute Content	Relative Changes of the Young's Modulus [40]		
	β	α	γ		Theory	Exp. T = 77 K	Exp. T = 300 K
Ti	−242	−191	−194	0.05 [40], 0.03 [88], 0.05 [86]	0.02	0.16	0.17
W	−198	−124	−147	0.03 [40]	0.05	0.13	0.15
V	−229	−182	−188	−0.03 [40], −0.02 [88]	0.04	0.07	0.08
Cr	−185	−175	−156	−0.05 [40], −0.02 [88], 0.01 [86]	0.02	0.02	0.02
Si	−227	−194	−231	-	0.03	0.09	0.08

Besides, the addition of low solubility alloying elements (e.g., Nb) may produce precipitates (Fe_2Nb). Niobium is one of the most effective alloying elements in increasing tensile strength at elevated temperatures. Precipitates, which are available in equilibrium in many of the Fe-Al-X systems, have a strong strengthening effect at high temperatures while they prevent grain coarsening but frequently lead to a considerable embrittlement at low or even ambient temperatures [37].

The effects of metallic substitutional atoms, and interstitials, such as boron [89] and carbon [90,91,92], on the mechanical and corrosion properties of Fe-Al alloys were also studied at different temperatures. Carbon produces plate-like κ-carbide (Fe_3AlC_x) precipitates which are predominantly located along dislocations and grain boundaries [35]. In the Ti-containing alloys, some isolated TiC precipitates were detected as well [35]. Boron could also make dispersoids with solute atoms. Borides can be dispersed in a Fe_3Al matrix after conventional casting methods [93]. Additionally, the thermal stability of Ti, Zr, Ta and Nb diboride precipitates in Fe_3Al-based material was proved earlier [93]. Ti, Hf and Zr diborides lead to significantly improved strengthening at elevated temperatures as well as higher creep resistances (Figure 11a [89]) compared to corresponding binary alloys. Interestingly, the existence of ZrB2 particles has been shown to enhance the strength at 600 °C and also the ductility at room temperature [94]. Maybe the grain structure refinement is the reason for the higher ductility of alloys.

Furthermore, some alloying elements could make stress induce phase transition or form hydride at low temperatures and in moist environments, which may lead to a cleavage-like fracture. The alloys most sensitive to the phase transition in the hydrogen environments are the Ti, Zr, Hf in the IVb column of the periodic table and the V, Nb and Ta in the Vb column of the periodic table. Figure 11b shows the effect of various alloying elements on the brittle to ductile transition temperature (BDTT). It is obvious that many alloying elements have a destructive influence on the ductility at low temperatures and enhance the BDTT temperature.

2.5.1. Influence of Solute Atoms on Mechanical Properties

Anthony and Fultz [97] proposed that the difference of the solute atomic radius compared to the Fe atoms controls their efficiency on the enhancement of transition temperature. However, the contribution of the alloying elements on the valence electron concentration may also play an important role for the transition temperature [88]. In a similar way, according to the classical theories of solid solution strengthening, the difference between solvent and solute atoms in atomic size (atomic size factor) produces local stress fields. It interacts with those of the dislocations, impeding their motion and causing an increase in the macro-hardness or flow stress. Additionally, the modulus misfit is due to the changes of binding energy of atoms after the addition of solute atoms. According to the model proposed by Fleischer [98,99] the magnitude of the solid solution strengthening can be expressed in terms of misfit parameters (ε) based on the following equation [100].

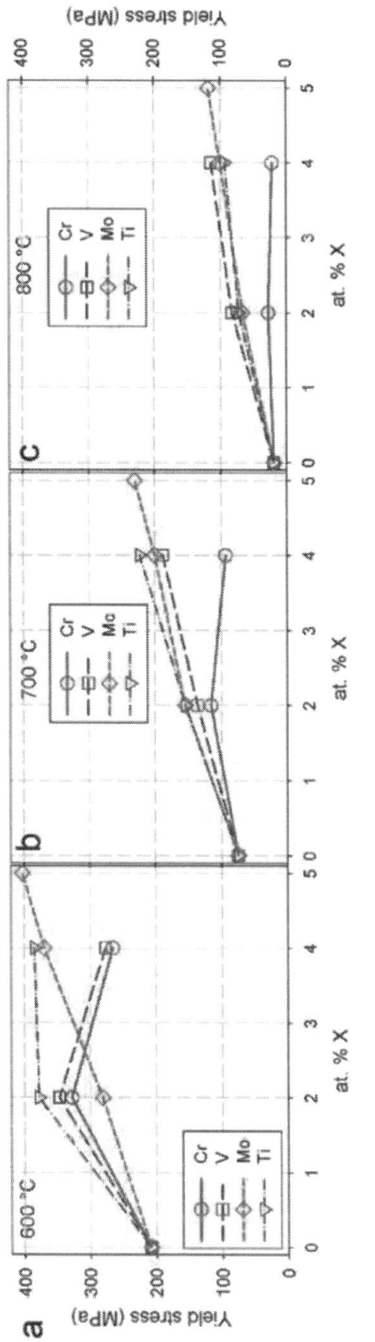

Figure 10. The effect of various ternary alloying elements on the yield stress at (**a**) 600 °C; (**b**) 700 °C; and (**c**) 800 °C [37].

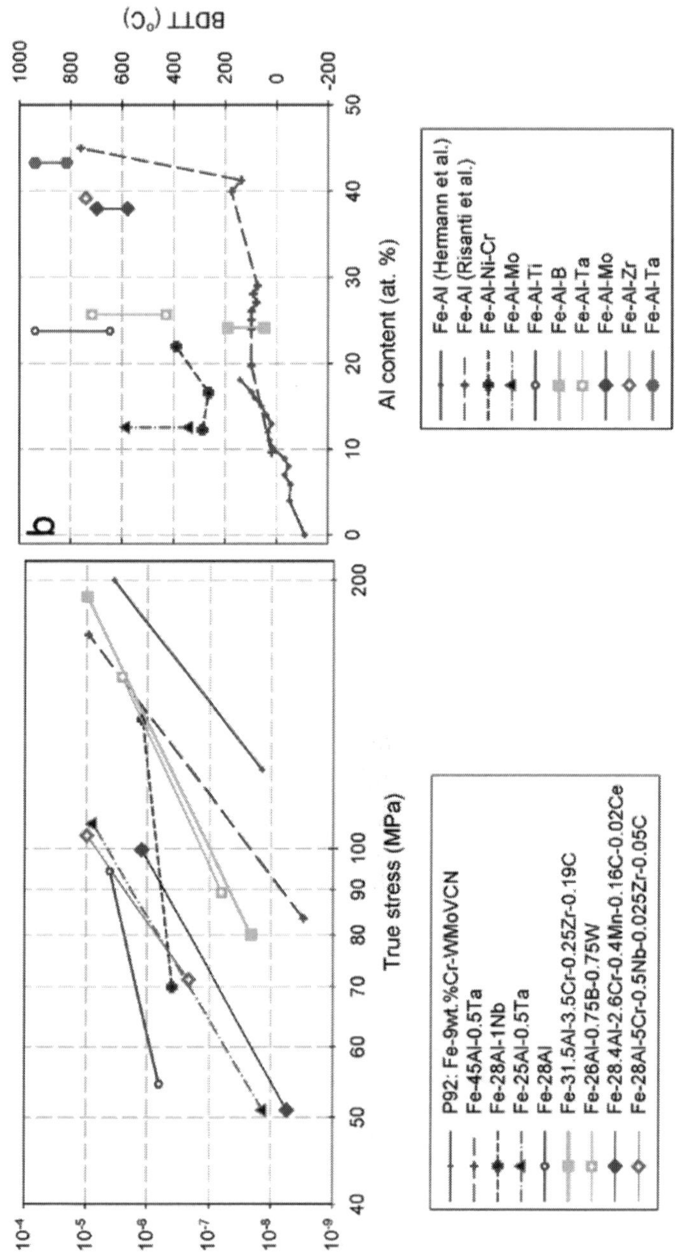

Figure 11. (**a**) Creep rate of various iron aluminides tested at 650 °C [89]; (**b**) Effect of various alloying elements on the brittle to ductile transition temperature (BDTT) [51,95,96].

$$\Delta\sigma = \eta M\mu\varepsilon^{1.5}C^{0.5} \quad (2)$$

where η is a constant, M is the Taylor factor and is constant, $\mu = \frac{E}{2(1+\upsilon)}$ is the shear modulus and C is the concentration of the additional element. The misfit parameters (ε) can be evaluated as follows [100];

$$\varepsilon = \left|\varepsilon'_{\mu} - \beta\varepsilon_a\right| = \left|\frac{\varepsilon_{\mu}}{1+0.5\left|\varepsilon_{\mu}\right|} - \beta\varepsilon_a\right| \quad (3)$$

where εμ=1μ/×ΔμΔC/ and εα=1a/×ΔaΔC/ are the modulus misfit parameter and atomic size factor, respectively. However, it is shown that the solid solution strengthening in intermetallics is more complicated compared with binary solutions [100]. For example, the sublattice that the solute atoms prefer to sit on controls the amount of the local stress field. The other complications are the contribution of alloying elements on the phase stability, grain size, energies of anti-phase boundaries and formation/migration enthalpy of the vacancies in B_2 or $D0_3$ structures. Hence, the precise characterization of thermal stability and/or solid solution strengthening in intermetallics requires the consideration of all the aforementioned factors.

2.5.2. Effect of Cr on the Mechanical Properties of Fe-Al Intermetallics

Cr is one of the most important alloying elements and its input on various thermo-mechanical and electrochemical properties has already been studied extensively [23,25,101,102,103,104]. Palm experimentally determined an isothermal section for the Fe-Al-Cr system at 1000 °C [82]. He did not find any ternary intermetallic phases because the binary phases can dissolve considerable amounts of the third component and the substitution does not make any changes in the crystallographic structure. Table 4 summarizes the effect of Cr on lattice constants and Young's modulus. It is shown that the measured Young's and shear modulus increase systematically with the increment of the Cr content in the alloys. The results were interpreted within the framework of the universal features of metallic bonding as a change in the interatomic potential, as proposed by Rose *et al.* [105,106]. The influence of Cr on ordering kinetics and dislocation configurations is not very significant [107]. Furthermore, the effect of Cr on the APBs is also still a matter of debate. While Kral *et al.* [74] found increased APB energy with Cr addition, McKamey *et al.* [25] found that Cr addition reduces the APB energy. Morris *et al.*[75] did not observe differences in the APB energies with Cr addition. Therefore, more experiments and simulations are needed for a precise conclusion about the Cr effect on APB energies.

The effect of Cr content on the incipient plasticity and post-yielding behavior of Fe-26Al-xCr alloys was studied with the aid of different local techniques such as nano-indentation as well as the punching and bending test of micro-pillars [108,109,110]. Strengthening of the interatomic bonds increases the dislocation line energy, and hence, enhances the energy needed for dislocation nucleation. Furthermore, it has been observed that the addition of 5 at. % Cr increases the nano-hardness at very low depths of indentation and also causes a higher flow stress (yield stress) at low strains in comparison to the binary alloys. This is due to the very low number of available dislocation sources (e.g., Frank-Read sources) in annealed materials [111] in addition to the higher energy needed for dislocation nucleation for Cr-rich alloys [110]. In contrast, at higher strains or depths of indentation, the binary alloy has higher hardness and flow stress (Figure 12). It shows that the amount of strain hardening in the ternary alloy is notably less than that in the binary alloy. It presents the sessile character of dislocations in the binary alloys and the lower ability of dislocations to cross-slip. Postmortem analysis of the nano-indents and punched pillars also proves the effect of Cr on the enhancement of the dislocation mobility, while the height of material pile-ups is decreased around the indents and the slip lines are wide spreading in the Cr-rich alloys [108].

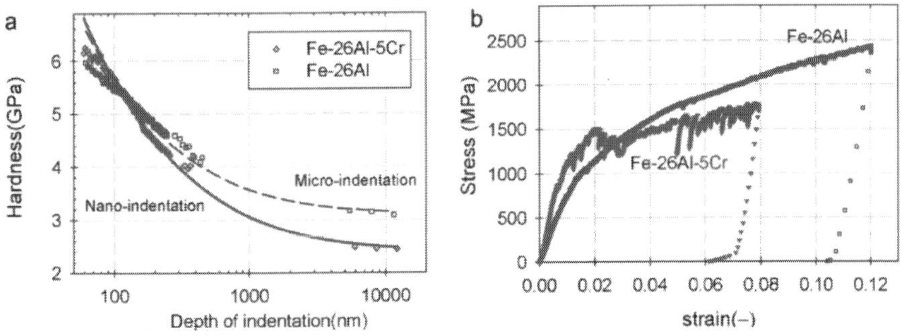

Figure 12. (**a**) The dependence of hardness on indentation depth and Cr content. Data from references [108,109]; (**b**) Stress-strain curves related to the Fe-26Al and Fe-26Al-5Cr micro-pillars. Data from reference [109].

The reduction of average values of material pile-ups after the addition of Cr (Figure 13) could be due to the increase of the capability of the material to include the plastic deformation, *i.e.*, how far dislocations are capable of transferring the plasticity into the material. This can be explained by considering the ease of dislocation cross-slipping in the Cr-rich samples. Additionally, very fine slip steps in the binary alloy (Figure 13e) related to the activation of several crystallographic slip planes were visible. In contrast, due to the localized slip on preferred glide planes, the glide steps of the Cr-containing alloy are well resolved on a few crystallographic planes (Figure 13f).

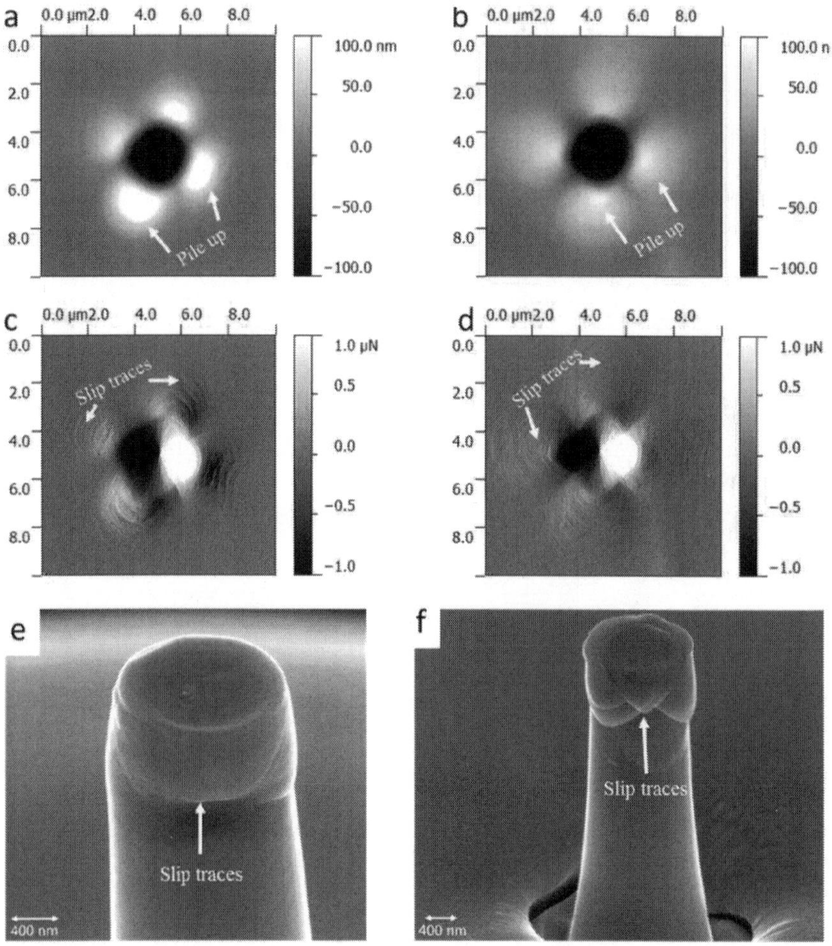

Figure 13. AFM topography and gradient images of material pile-ups for Fe-26Al (**a**) and (**c**) and Fe-26Al-5Cr (**b**) and (**d**). Data from reference [108]. (**e**) and (**f**) represent the slip traces after ~10% strain of Fe-26Al and Fe-26Al-5Cr micro-pillars. Data from reference [109].

3. ENVIRONMENTAL DEGRADATION

3.1. Oxidation and Corrosion of Iron-Aluminides

In a binary alloy, one constituent may segregate to the surface while the surface energy is mostly lower than the bulk-terminated alloy surface [112]. Experimental studies on clean, low-index Fe_{1-}

$_xAl_x$ surfaces show the segregation of Al atoms at the surface. The higher the Al content the more Al exists at the surface. Higher temperatures enhance the Al segregation even more [113,114]. For a stoichiometric Fe_3Al with (110) normal orientation, for example, Al segregation at room temperature was seen to be about 41 at. %, while at 427 °C (below the $D0_3$-B_2 phase transition) the Al content is about 94 at. % within the first layer [115]. It has been shown that the top three layers of atoms may be strongly influenced by surface segregation phenomena and may have a complicated structure [116]. The deviation of the atomic concentration at the surface compared to the bulk material along with the surface energies causes different electronic, chemical and physical properties at the top surface of the samples.

3.1.1. Oxidation Resistance

The Fe-Al intermetallics are aimed to be used in hostile environments, and hence, studying the environmental aspects on the chemical and mechanical properties of alloys is essential. In general, the oxide growth mechanism consists of two steps; (i) adsorption of oxygen on the metal surface and (ii) formation of an oxide layer, if the coverage of the metallic surface with the oxide layer is energetically favored and no kinetic barriers exist. Once a first oxide layer has formed, it grows further in thickness. Within this second step a charge transfer through the oxide scale is crucial to preserve charge neutrality and therefore further growth. In the studies of the oxidation behavior of B2 aluminides, a parabolic oxide growth rate was observed at high temperatures [117]. It occurs mostly by oxygen diffusing inward through the oxide, with oxide formation taking place at the metal/oxide interface.

The chemical composition of the oxide layer strongly depends on the (i) oxidation rate; (ii) the amount of the alloying elements at the surface, (iii) temperature; (iv) environment or oxidizing media; and (v) the presence of kinetic barriers at the surface. Formation of aluminum oxide is thermodynamically more favorable than iron oxide since oxygen has a higher affinity to Al compared to Fe (the standard Gibbs energy of formation of α-Al_2O_3 is −1,582,260 J/mol which is almost 6.5 times higher compared to FeO). However, the experimental results performed on binary polycrystals (with 15 and 40 at. % Al), oxidized at temperatures around 727 °C in synthetic air, show the existence of a thin Fe_2O_3-containing outer layer followed by an inner, nearly pure Al_2O_3 layer. Additionally, a complex oxide formation (e.g., $AlFeO_3$ or $FeAl_2O_4$ = FeO. Al_2O_3) may happen at low segregation rates of Al from the bulk or high oxidation rates. However, annealing the aluminides at about 1000 °C after oxidation breaks up bonds between Fe and O and leads to the formation of pure Al oxide. In general, at the very beginning of the oxidation process, the oxidation of Fe atoms may happen. After the formation of homogeneous mixed Fe and Al oxides, the supply of oxygen to the metal-oxide interface allows selective oxidation of Al and thereby the development of a continuous film of alumina. The formation of a homogeneous Al_2O_3 layer further slows down oxide growth because electron tunneling remains the only possibility for charge equalization. It may block the Fe oxide formation. Therefore, the existence of a large amount of Al within the surface region in addition to low O_2 pressures in the oxidizing atmosphere causes the formation of pure alumina films. The alumina oxide layer can be either metastable γ-, θ-Al_2O_3, or the slowest-growing, thermodynamically stable α-Al_2O_3, known as corundum. The metastable phases may form at low temperatures (about 700 °C) and transform to the α-Al_2O_3 at high temperatures or after some time. In short, the Fe_2O_3content decreases with time, temperature and Al concentration of the bulk alloy [118].

The α-Al_2O_3 maintains its desirable oxidation resistance over a wide range of temperatures and in steam environments [119,120,121, 122,123]. Increasing the Al concentration enhances the protectivity of the oxide layer at high temperatures [17,18]. A bulk sample with about 14 and 19 at. % Al could maintain its protectivity at temperatures above 800 and 900 °C, respectively [124]. Oxidation resistance of aluminides can be improved by the additions of rare earth metals and transition metals such as Cr, Ti, Nb [37], Y and Ce [125]. However, the addition of high concentrations of Mo or Zr may increase the corrosion rate of iron aluminides (Figure 14) [37]. Additionally, at temperatures higher than 1000 °C, Cr_2O_3 is unstable and its evaporation will increase the oxidation rate of alloys. The addition of small amounts of reactive elements, e.g., Y, Ce, Hf to chromia- and alumina, improves the adhesion of the oxide to the substrate. The presence of alloying elements may limit the outward diffusion of aluminum ions in the oxide layer and, hence, the oxide grows by the inward diffusion of oxygen [126]. On the other hand, alloying elements in the precipitated state may deteriorate the oxidation behavior in terms of internal oxidation. The precipitate-matrix interfaces catalyze the formation of Al_2O_3 and act as starting points for internal oxidation. They are fast paths for oxygen inward and aluminum outward diffusion [124].

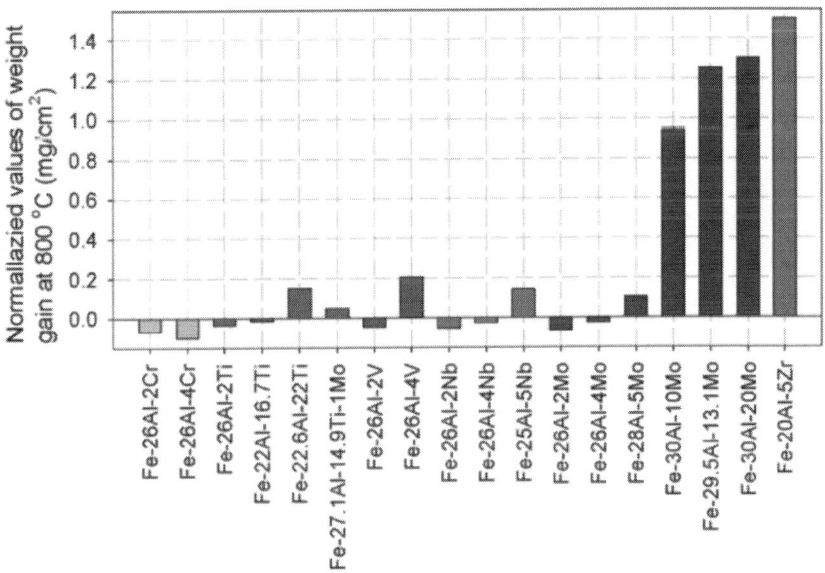

Figure 14. Normalized values of weight gains of Fe-Al-based intermetallics in air (zero line is the mass gain of Fe-26Al) [37].

3.1.2. Electrochemical Properties of Fe-Al Intermetallics

Resistance to aqueous corrosion at low (room) temperatures is essential for the applicability of iron aluminides without compromising their structural integrity. Remarkable enrichment of Al within the passive film [24,127] plays a beneficial role for the passivation of Fe-Al intermetallics. A Pourbaix diagram of aluminum [128] provides information about the thermodynamic stability of different species as a function of potential and pH (Figure 15). At room temperatures and in near-neutral and Cl^--free solutions, aluminum provides a protective insoluble oxide/hydroxide film [129,130,131]. However, the range of passivity varies with temperature, presence of secondary phases and/or precipitates (in bulk material and/or the oxide layer) and existence of substances in the solution that can form soluble complexes or insoluble salts with aluminum [132,133].

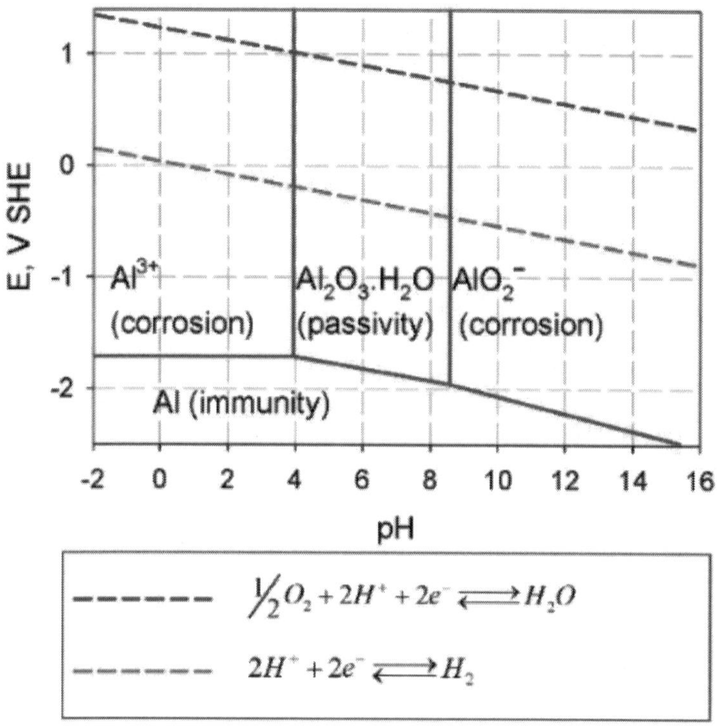

Figure 15. Pourbaix diagram for aluminum at 25 °C.

The contributions of alloying elements on the electrochemical properties of iron aluminides in aqueous environments have rarely been studied. Electrochemical impedance spectroscopy (EIS) and the

Mott–Schottky (MS) methods were performed to characterize the passive film of binary and ternary alloys with different Cr content in the near-neutral and Cl^--free solutions [101]. The existence of high donor densities in the passive layers [101] causes semi-conductive behavior. In the sample with a higher concentration of chromium, some part of the iron oxide in the passive layer was replaced by Cr^{3+}Contribution of Cr ions (Cr^{3+} and Cr^{6+}) to the increase of the effective capacitance and donor density of the passive layers was seen based on the electrochemical measurements. The effect of Cr on the polarization resistance was more obvious at high anodic potentials where the strong enrichment of the passive layer with Cr^{3+} and Cr^{6+} could make a more protective p-type passive layer, instead of the n-type (Al^{3+}-rich) layer, for example at lower potentials or in alloys without Cr [101]. Additionally, it has been shown that the reduction of the flat-band potential after the addition of 5 at. % Cr may reduce the sensitivity of alloys to the moist-induced hydrogen embrittlement [101].

Like other commercial Al-rich alloys, iron-aluminum intermetallics have poor resistance to localized (pitting and crevice) corrosion in Cl^--containing solutions [101,134] (Figure 16 [101]). The incorporation of Cr as a ternary alloying element into the passive film increases the range of passivity and, consequently, the resistance of the alloys to pitting corrosion and decreases the average density of the pits. However, the addition of 5 at. % Cr to the binary Fe_3Al intermetallic does not completely hinder the pitting and crevice corrosion of the alloys. Moreover, the beneficial effect of Si and Ge on the passivating characteristics of Fe_3Al intermetallics has been observed [135]. Zamanzade *et al.* [101] tried to characterize the form and density of (meta) stable pits for different crystal orientations. It was shown that the pit facets have the {110} orientation. The grains with the {110} orientation have slightly higher pit density in comparison with other orientations, but the difference was not significant [101].

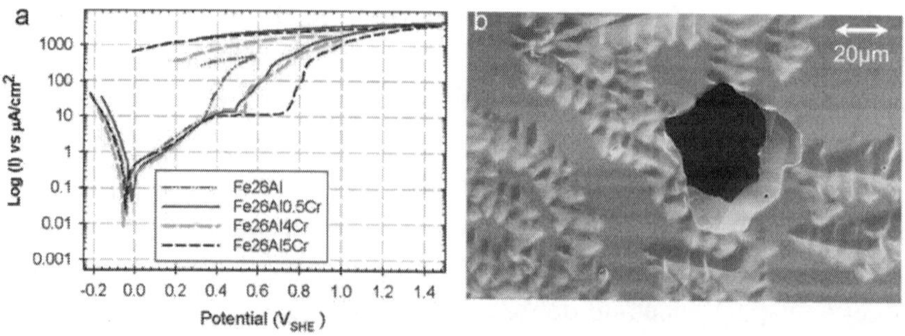

Figure 16. (**a**) Polarization curves show the effect of Cr on the increase of passivation range in a Cl^--containing solution; (**b**) SEM image shows the formation of micro-pits in the iron aluminides. Data from reference [101].

According to the Pourbaix diagram, the corrosion rate increases as the pH moves away from the near-neutral condition. In many acidic solutions, Al dissolves as Al^{3+} ions. In alkaline environments, aluminum dissolves as AlO_2^-. Moreover, like pure aluminum, dissolution of Al_2O_3 may happen due to local alkalization induced by the hydrogen evolution reaction at high cathodic current densities. Al (oxide) dissolution reaction under hydrogen evolution is described as follows [24]:

$$Al_2O_3 + 3H_2O + 2OH^- \rightarrow 2Al(OH)_4^- \quad (4)$$

$$Al + 4OH^- \rightarrow Al(OH)_4^- + 3e^- \quad (5)$$

3.2. Hydrogen Embrittlement (HE)

Since Johnson [136] first reported on hydrogen embrittlement, various strong views on the mechanisms of hydrogen embrittlement have been vigorously discussed and thoroughly reviewed in the literature. It was shown that hydrogen embrittlement causes severe mechanical degradation of various materials [137,138], especially in aluminides [25,139]. In iron-aluminides, the environmental embrittlement was considered a major cause for the low ductility at ambient temperatures and in air [16,25]. High reactivity of Al atoms with the moisture in air creates hydrogen, and it is the resulting hydrogen atoms that are then responsible for the lowered ductility. At crack tips, this reaction results in embrittlement due to the formation of the atomic hydrogen, which penetrates the material in front of the crack tips.

3.2.1. Mechanism of Hydrogen Ingress

The corrosion of aluminum (Equation (6)) originates the water reduction reaction (Equation (7)) in aqueous environments or moist atmospheres. The consequence of the electro-chemical reactions can be written as follows:

$$Al \rightarrow 3\overline{e} + Al^{3+} \tag{6}$$

$$3 \times \left[H_2O + Al + \overline{e} \rightarrow AlH_{ads} + OH^-\right] \tag{7}$$

Next, hydrogen is adsorbed (H_{ads}) at the metallic surface as atomic hydrogen. Then, a part of the atomic hydrogen recombines chemically (Tafel reaction (Equation (8)) or electrochemically (Heyrovsky mechanism (Equation (9)) to form molecular hydrogen, H_2, which leaves the metallic surface [140]. The molecular hydrogen does not cause severe embrittlement, possibly as a result of its lower activity in comparison to the atomic hydrogen.

$$AlH_{ads} + AlH_{ads} \rightarrow 2Al + H_2 \tag{8}$$

$$AlH_{ads} + H_2O + e^- \rightarrow Al + H_2 + OH^- \quad (9)$$

A part of H_{ads} will undergo an absorption (H_{abs}) reaction inside the material (Hads⇔Habs). The passage of atomic hydrogen through the alloy/solution interface depends on the surface coverage (θ) and also the number of available sites in the subsurface that the hydrogen can occupy. The consequence is the accumulation of hydrogen under the interface, leading to a concentration C_0. Density functional theory's local density approximation calculations [141], in addition to periodic density functional theory calculations within the generalized gradient approximation [142], predicted that hydrogen sits at tetrahedral sites in the bulk Fe-Al lattice, like its interstitial sites in bulk Fe [143]. The activation barrier of some different alloys is presented in Table 5 [142,144,145]. Hydrogen diffusivity in iron-aluminides is lower than in pure Fe (Table 5) [144,146,147]. Electrochemical permeation tests [148,149,150] found that hydrogen diffusivity in Fe-Al alloys decreases with increasing Al content [146,147,148,149,151]. However, new experimental diffusivity data for Fe-Al alloys reported by [149] vary significantly from the older data, presumably due to experimental uncertainties, varying Al concentration, impurities and microstructural differences [142]. Increasing the Cr content of iron [152] or iron aluminum [146] decreases the diffusion coefficient and increases the solubility of hydrogen since the heat of absorption decreases and d-vacancies increase with substitution of Fe with Cr [152].

Table 5. The activation barrier and diffusion coefficient of some different alloys.

Material	Activation Barrier (eV)	Diffusion Coefficient at Room Temperature (m^2/s)
Fe	–	10^{-8}
Fe-18Al	–	10^{-11} [146,148]
Fe-25Al	0.42 [144]	1.45×10^{-13} [146]
Fe-37Al	–	5.57×10^{-10} [149]
Fe-40Al	0.22 [145]	4.4×10^{-13} [147], 5.07×10^{-10} [149]
Fe-43Al	–	4.46×10^{-10} [149]
Fe-46Al	–	3.62×10^{-10} [149]
Fe-50Al	0.26 [142]	2.257×10^{-10} [149]

For the passivated materials, the transport of hydrogen through the passive layer is driven by the electric potential gradient and the hydrogen concentration gradient [153] in contrast to the transport of hydrogen in bulk materials, which is controlled mainly by the concentration gradients. The ionic characteristic of hydrogen atoms (protons) in oxides [154] makes a strong columbic interaction between hydrogen-oxygen ions. It may cause a very high concentration of hydrogen in the oxide in comparison with bulk metal or the metal/oxide interface. For Fe, the ratio of hydrogen concentration in oxide to bulk is measured to be approximately 10^6 [153]. Additionally, the necessity for the breaking of the columbic bond with the initial oxygen ion causes the mobility of hydrogen in an oxide to be considerably lower than that in the metal phase.

The transport of hydrogen within the oxide must occur by the activated jumping of a proton from one oxygen ion to another [142]. In the case of Fe, for example, the hydrogen diffusivity in the iron oxide layer is much less than that in iron (around 10^{-14} m^2/s) [153,155]. Based on the density functional theory simulation [156], existence of an Al_2O_3 layer on the Fe-Al bulk material suppresses the mobility of hydrogen atoms. Diffusion of the adsorbed hydrogen atoms from the outer part of the solution/oxide layer into the α-Al_2O_3 layer is thermodynamically unfavorable, endothermic and rate-controlling. In contrast, the H diffusion from the inner part of α-Al_2O_3 to the oxide/bulk interface, as well as from the interface into the bulk of the Fe-Al, are thermodynamically spontaneous and exothermic processes [156]. It should be mentioned here that the diffusion mechanism through the passive layer and also the passive layer/bulk metal interface is still largely unknown and has, so far, not been completely explored in published papers. Due to the existence of high donor densities in the passive layers [101], the real thermodynamics and kinetics of hydrogen diffusion may be far from the predicted values based on simulations or the measurements performed on aluminum-rich coatings [157]. The semi-conductive properties of the oxide layer result in a rectification effect, which is the easy movement of current carriers (*i.e.*, electron and protons, H^+) from the oxide/electrolyte towards the metal/oxide interface. It eases the reaction of the absorbed H^+ ions with the Al substrate at defect sites, which follows with the oxidization of the Al and reduction of H^+ [158]. Thus, an increase in hydrogen diffusivity can be expected with an increase in the density of defects and distortion of the long-range order of oxide film.

3.2.2. Hydrogen Interaction with Defects

Along with the conventional hydrogen solubility in the lattice matrix, hydrogen congregates in lattice defects with large surface areas, such as grain boundaries, dislocations, and voids. Point defects such as monovacancies, interstitial atoms and other impurity atoms can also act as hydrogen-trapping defects [159]. The traps are divided into reversible and irreversible or low and high energies, respectively. Table 6 [160,161,162,163,164] shows the strong binding energy of hydrogen at sites near the dislocation core, vacancies and free surfaces.

Table 6. Binding energies of various traps in Fe and Al.

Host Material	Type of Trap	Binding Energy (eV)	Reference (s)
α-Fe	Vacancy	0.63, 0.48	[160,161]
	substitutional (Ti)	0.19	[160]
	interstitional (C)	0.03	[160]
	interstitional (N)	>0.13	[160]
	Grain boundary	0.10	[161]
	Dislocation elastic stress field	0.21	[161]
	Dislocation core	0.61	[161]
	Free surface	0.73	[161]
bcc carbon steel	Fe_3C phase interface	0.11	[160]
Al	Vacancy	0.52, 0.53	[160–164]
	Grain boundary	0.15	[160]
	Al_2O_3-Al phase interface	0.7, 1.0–1.4	[160]

The total hydrogen content in a metal at a defined temperature (C_T^H) is the sum of the solute hydrogen atoms (C_L^H) and the trapped atoms at different defects [143].

$$C_T^H = C_L^H + \sum_j k^H n_j^H N_j^H \quad (10)$$

where k^H is a constant, n_j^H and N_j^H are the fraction of trap sites filled with hydrogen and the total number of trap sites per unit volume, respectively. For the dislocations $N_{dis}^H \approx \pi b^{-1}\rho$, where b and ρ are the burgers vector and dislocation density, respectively. For the grain boundaries, the incomplete information about the atomic arrangement may cause an error in the estimated hydrogen binding energies. However, as a first estimation, we can assume the total number of trap sites per unit volume as $N_{gb}^H \approx b^{-2}L_{gb}$, where L_{gb} is the length of grain boundaries per unit area of observation. Due to the small volume fraction of the metal which is disturbed by the defect, quantification of the interactions between defect and hydrogen may be very difficult. According to Hirth [143], the trapping effect is measureable for a trap density of 10^{23} m^{-3} or more, and is negligible for smaller trap densities at temperatures below 550 K. Dislocation trap densities can exceed this value at room temperature, while recovery decreases the dislocation densities.

The existence of traps also influences the hydrogen diffusion coefficient in different ways. Substitutionally dissolved impurities accelerate the hydrogen diffusion [160], which relates to reduced vacancy formation energy in the excess volume of the boundary. In contrast, the hydrogen diffusion does not rely on vacancies [160]. The mobility of hydrogen can be substantially reduced by its attractive interactions with dislocations. On the other hand, for high purity single-phase single crystals the diffusion paths are either bulk diffusion or pipe diffusion along dislocations [165]. It is possible that the hydrogen does not enter the lattice by diffusion, but is transported into the lattice by gliding dislocations that originate at the surface. The "Cottrell atmosphere", formed around dislocations, could simultaneously move with the dislocation during deformation and improve the transport of hydrogen, if the velocity of the dislocations is less than a critical value [166,167,168]. The critical velocity (v_c) could be analytically calculated based on the following equation [167]:

$$v_c = \frac{\sigma b l D_{ok}}{RT} \exp\left(-\frac{Q_k + 2F_k^*}{RT}\right) \tag{11}$$

where σ, *l*, *R*, *T*, D_{ok}, Q_k and $2F_k^*$ are applied stress, dislocation length, gas constant, absolute temperature, pre-exponential term for kink diffusion, activation energy for kink diffusion and free energy of formation of a double-kink on a dislocation, respectively. The interaction between hydrogen and dislocations in metals influences not only the hydrogen content and its mobility but also the dislocation motion and plastic flow of material. Hydrogen may suppress the motion of dislocation which reflects the endothermic detrapping necessary to separate the moving dislocations from its hydrogen or, contrarily, the dislocation transport may ease with hydrogen, based on the hydrogen-enhanced local plasticity mechanism.

3.2.3. Hydrogen Embrittlement of Fe-Al Intermetallics

Because of the technological importance of hydrogen embrittlement, many people have explored the nature, causes and control of hydrogen-related degradation of aluminides. There are at least three reasons for the high susceptibility of iron aluminides to hydrogen embrittlement: (i) high reactivity of Al atoms with the moisture in air creates high fugacity of hydrogen atoms; (ii) existence of a high concentration of point defects and especially vacancies (Figure 6b) enhances the solubility of hydrogen in the intermetallics, and increases the influence of hydrogen on the mechanical properties; (iii) the iron aluminides intrinsically have a limited amount of ductility (and very high slip planarity) in comparison to bcc metals, and an additional reduction with hydrogen charging will decrease the ductility even more. The effect of hydrogen on the ductility and crack initiation of iron aluminides was a subject of many studies [169,170,171,172]. The polycrystalline Fe-36.5Al was shown to have ductility of just 2.2% in air and 5.4% in vacuum while its ductility in dry oxygen could be as high as 17.6% [16] (Figure 17). However, the fracture of Fe-Al polycrystalline materials seems to be dominated by the intergranular crack growth mechanism in moist environments, while the cracks will grow transgranularly in the Fe_3Al structures. For single crystals of Fe-40Al tested in air, crack initiation occurred at the surface and the fracture strain was less than 1%. Specimens tested in an O_2 atmosphere had internally initiated cracks, possibly at flaws, and elongations of ~10% [169,170,171,172]. Wittmann *et al.* [165] studied the ductility of Fe-40Al single-slip-oriented single crystals. They also observed improved ductility when tested in vacuum *versus* air, but greater ductility in air (between 1.1%–5.9%) compared with previously reported results for multiple slip-oriented specimens. The elongations and fracture strengths of Fe-40Al single crystals are shown in Table 7 [165]. The scatter, shown in the results, is believed to be due to the small thickness of the tensile specimens. However, the ductility appears to be less in air than in a vacuum. Additionally, higher fracture strengths were seen in a vacuum. Wittmann *et al.* [165] did not observe any differences in environmental embrittlement of specimens oriented in such a way that the strain was predominately carried by screw dislocations or edge dislocations. On the contrary, results from tensile tests of Fe_3Al single crystals, oriented in a way that the main strain was produced either by screw dislocations or by edge dislocations, have shown considerable ductility for the screw orientation in both air and vacuum, whereas the edge-oriented specimens showed little ductility in air [171]. Saka and Nishizaki [171] related this difference to the role of edge dislocations in transporting hydrogen into the Fe_3Al lattice. The crack in FeAl propagates mainly along the {100} plane in air [169,170]. In contrast, in a vacuum, fracture has been shown to occur along {111} for stoichiometric Fe-Al, and {100} for Fe-40Al, and Fe-35Al [169,170]. Munroe and Baker [173] suggested a concept for HE of iron aluminides. They proposed that the {100} cleavage that results from the interaction of two a/2⟨111⟩ edge super-partials may produce a sessile ⟨100⟩ edge dislocation on {001}. This edge dislocation can then act as a crack nucleus. Moreover, formation of the ⟨100⟩ edge dislocations may enhance absorption of hydrogen rather than two super-partials [174].

At a constant temperature, and in the absence of any kind of phase transformation (see Section 2.5), the energy of the newly formed surfaces (due to the crack growth) is an intrinsic characteristic of the material. The hydrogen ingress into the materials can reduce the energy needed for formation of new surfaces with decrement of the strength of interatomic bonds, as predicted by the hydrogen-enhanced decohesion (HEDE) mechanism. On the other hand, the presence of hydrogen affects the multiplication or movement of dislocations in agreement with the DEFect ACTing AgeNTS (defactants), adsorption-induced dislocation emission (AIDE), hydrogen-enhanced local plasticity (HELP), *etc.*, theories [24].

However, the magnitude of the influence of hydrogen on each item depends on the alloy content, defect concentrations, temperature, environment, *etc.* In order to clarify the most probable mechanism of hydrogen embrittlement for each material in a defined environmental situation, it is important to resolve the impact of hydrogen separately on each of the mentioned items, which is impossible based on the conventional mechanical testing methods. Hence, the results of macro-mechanical tests present, sometimes, complicated and controversial findings and/or interpretations. In contrast, local techniques enable us to measure the influence of hydrogen on elastic properties, pop-in load and hardness or flow stress, which scale with the strength of interatomic bonds, energy needed for homogeneous dislocation nucleation and mobility of dislocations, respectively [22,24,29]. A very slight reduction of Young's modulus was measured experimentally due to hydrogen charging for intermetallics with different Cr content (Figure 18a) [24]. A 3%–5% reduction of Young's modulus could not explain the high susceptibility of alloys to HE. Moreover, a strong reduction of the energy needed for homogeneous dislocation nucleation (HDN) after hydrogen charging was seen in the binary alloys (Figure 18b [24]). It is probably the main reason for the very high sensitivity of binary alloys to hydrogen embrittlement. The addition of chromium decreases the susceptibility of binary alloys to the HE because the dislocation core radius does not change significantly after hydrogen charging [24]. The increase of the hardness (Figure 18c,d) of alloys after hydrogen charging may be due to the increase of the friction stress of dislocations. All this experimental evidence shows that adsorption of hydrogen will ease dislocation nucleation while the dislocations are less glissile. Therefore, it is proposed that the mechanism of dislocation shielding should be considered for analyzing the fracture characteristics of iron aluminides in atmospheres containing hydrogen [24].

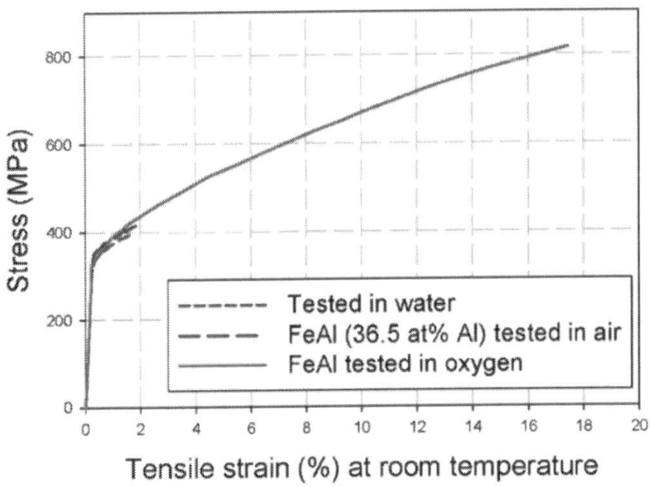

Figure 17. Stress-strain curves of Fe-Al intermetallics in different environments at room temperature. Data from reference [16].

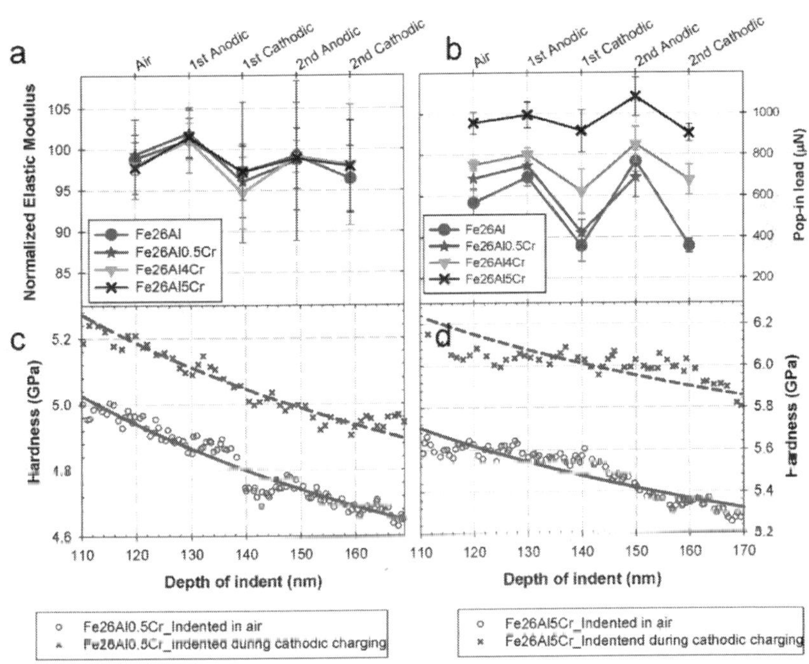

Figure 18. The effect of hydrogen on Young's modulus (**a**); pop-in load (**b**); and hardness (**c** and **d**) for Fe-26Al-xCr alloys. Data from reference [24].

Table 7. Elongation and fracture strength of Fe-40Al in different environments after performing tensile test [165].

Sample Orientation	Elongation (%)		Fracture Strength (MPa)	
	Air	Vacuum	Air	Vacuum
Edge	4.0	9.4	336	460
	3.3	5.7	335	425
	5.9	12.0	320	422
Screw	1.1	3.0	235	435
	2.6	10.8	335	495
	5.7	10.1	345	470

4. CONCLUDING REMARKS

This review paper aims to improve the understanding of the factors that control the microstructure and properties of iron aluminides. Some thermo-mechanical behaviors of iron aluminides were studied and compared with super-alloys and stainless steels. Additionally, current applications of the alloys were summarized. The influence of alloying elements on the phase transformation, mechanical properties, corrosion and hydrogen embrittlement were studied. Moreover, the most important barriers for the wide usage of iron aluminides were considered. They include (i) hydrogen embrittlement in the moisture-containing atmospheres at room temperature; (ii) localized corrosion in the Cl^{-}-containing aqueous solutions; (iii) strong reduction of yield stress after a critical temperature; (iv) limited creep resistance; and (v) high thermal conductivity. Finally, possible ways for overcoming these problems were evaluated.

ACKNOWLEDGMENTS

The authors wish to thank Horst Vehoff, Jorge Rafael Velayarce and Michael Nackas for their advice and help and the Deutsche Forschungsgemeinschaft (DFG) for financial support.

AUTHOR CONTRIBUTIONS

All authors helped conduct the literature review and prepare the text. Mohammad Zamanzade and Afrooz Barnoush performed electrochemical tests as well as *ex-situ* and *in-situ* nano-indentation measurements. All authors helped to accomplish the tensile tests and micro punching tests of iron aluminides.

CONFLICTS OF INTEREST

The authors declare no conflict of interest.

REFERENCES

1 Ziegler, N. Resistance of iron-aluminum alloys to oxidation at high temperatures. Trans. Am. Inst. Min. Met. Eng. 1932,100, 267–271.

2 Blau, P.J.; Meyer, H.M. Characteristics of wear particles produced during friction tests of conventional and unconventional disc brake materials. Wear 2003, 255, 1261–1269.

3 Judkins, R.R.; Rao, U.S. Fossil energy applications of intermetallic alloys. Intermetallics 2000, 8, 1347–1354.

4 Stoloff, N.; Liu, C.; Deevi, S. Emerging applications of intermetallics. Intermetallics 2000, 8, 1313–1320.

5 Baker, I.; Munroe, P.R. Mechanical properties of FeAl. Int. Mater. Rev. 1997, 42, 181–205.

6 Noebe, R.D.; Bowman, R.R.; Nathal, M.V. Physical and mechanical-properties of the B2 compound nial. Int. Mater. Rev. 1993, 38, 193–232.

7 Sundararajan, V.; Sahu, B.R.; Kanhere, D.G.; Panat, P.V.; Das, G.P. Cohesive, electronic and magnetic-properties of the transition-metal aluminides FeAl, CoAl and NiAl. J. Phys.Condens. Matter 1995, 7, 6019–6034.

8 Fu, C.L.; Wang, X.D. The effect of electronic structure on the defect properties of FeAl. Mat. Sci. Eng. A-Struct. 1997,240, 761–768.

9 Fu, C.L.; Yoo, M.H. Electronic structure and mechanical behavior of transition-metal aluminides: A first-principles total-energy investigation. Mater. Chem. Phys. 1992, 32, 25–36.

10 Miracle, D.; Darolia, R. NiAl and Its Alloys, Intermetallic Compounds, Vol. 3: Structural Applications of Intermetallic Compounds; Wiley: New York, NY, USA, 1994.

11 Huang, S.C.; Chesnutt, C.J. Gamma TiAl and Its Alloys, Intermetallic Compounds, Vol. 3: Structural Applications of Intermetallic Compounds; Wiley: New York, NY, USA, 1994.

12 Briant, C.L. Intergranular and Cleavage Fracture, Intermetallic Compounds, Vol. 2: Basic Mechanical Properties and Lattice Defects of Intermetallic Compounds; Wiley: New York, NY, USA, 1994.

13 Vedula, K. FeAl and Fe3Al, Intermetallic Compounds, Vol. 3: Structural Applications of Intermetallic Compounds; Wiley: New York, NY, USA, 1994.

14 Mckamey, C.G.; Devan, J.H.; Tortorelli, P.F.; Sikka, V.K. A review of recent developments in Fe3Al-based alloys. J. Mater. Res. 1991, 6, 1779–1805.

15 Liu, C.T. Recent advances in ordered intermetallics. Mater. Chem. Phys. 1995, 42, 77–86.

16 Liu, C.; Lee, E.; McKamey, C. An environmental effect as the major cause for room-temperature embrittlement in FeAl. Scr. Metall. Mater. 1989, 23, 875–880.

17 Prescott, R.; Graham, M. The oxidation of iron-aluminum alloys. Oxid. Metals 1992, 38, 73–87.

18 Tortorelli, P.; Natesan, K. Critical factors affecting the high-temperature corrosion performance of iron aluminides.Mater. Sci. Eng.: A 1998, 258, 115–125.

19 Alven, D.; Stoloff, N. The influence of composition on the environmental embrittlement of Fe3Al alloys. Mater. Sci. Eng.: A 1997, 239, 362–368.

20 Castagna, A.; Stoloff, N. Hydrogen embrittlement of Fe3Al alloys. Mater. Sci. Eng. A 1995, 192, 399–406.

21 Balasubramaniam, R. Hydrogen in iron aluminides. J. Alloys Compd. 2002, 330, 506–510.

22 Barnoush, A.; Dake, J.; Kheradmand, N.; Vehoff, H. Examination of hydrogen embrittlement in FeAl by means of in situ electrochemical micropillar compression and nanoindentation techniques. Intermetallics 2010, 18, 1385–1389.

23 Zamanzade, M.; Barnoush, A. An overview of the hydrogen embrittlement of iron aluminides. Procedia Mater. Sci.2014, 3, 2016–2023.

24 Zamanzade, M.; Vehoff, H.; Barnoush, A. Cr effect on hydrogen embrittlement of Fe3Al-based iron aluminide intermetallics: Surface or bulk effect. Acta Mater. 2014, 69, 210–223.

25 McKamey, C.; Horton, J.; Liu, C. Effect of chromium on properties of Fe3Al. J. Mater. Res. 1989, 4, 1156–1163.

26 Keddam, M.; Mattos, O.; Takenouti, H. Mechanism of anodic dissolution of iron-chromium alloys investigated by electrode impedances—I. Experimental results and reaction model. Electrochim. Acta 1986, 31, 1147–1158.

27 Keddam, M.; Mattos, O.; Takenouti, H. Mechanism of anodic dissolution of iron-chromium alloys investigated by electrode impedances II. Elaboration of the reaction model. Electrochim. Acta 1986, 31, 1159–1165.

28 Epelboin, I.; Keddam, M.; Mattos, O.; Takenouti, H. The dissolution and passivation of Fe and Fe-Cr alloys in acidified sulphate medium: Influences of pH and Cr content. Corros. Sci. 1979, 19, 1105–1112.

29 Barnoush, A.; Vehoff, H. Recent developments in the study of hydrogen embrittlement: Hydrogen effect on dislocation nucleation. Acta Mater. 2010, 58, 5274–5285.

30 McQueen, H.J.; Kuczynski, G.C. Order-disorder transformations in iron-aluminum alloys. Trans. Am. Inst. Min. Metall. Eng. 1959, 215, 619–622.

31 Eguchi, T.; Matsuda, H.; Oki, K.; Kiyoto, S.-i.; Yasutake, K. Order-disorder transformation in Fe-Al alloys. Trans. Jpn. Inst. Metals 1967, 8, 174–179.

32 Koster, W.; Godecke, T. Physical measurements on iron—Aluminum alloys between 10 and 50 at.% Al. I.—Confirmation of and additional conribution to the iron—Aluminum phase diagram. Z. Metallkd. 1980, 71, 765–769.

33 Taylor, A.; Jones, R. Constitution and magnetic properties of iron-rich iron-aluminum alloys. J. Phys. Chem. Solids1958, 6, 16–37.

34 Davies, R. An X-ray and dilatometric study of order and the "k-state" in iron-aluminum alloys. J. Phys. Chem. Solids1963, 24, 985–992.

35 Stein, F.; Schneider, A.; Frommeyer, G. Flow stress anomaly and order–disorder transitions in Fe3Al-based Fe-Al-Ti-X alloys with X = V, Cr, Nb, or Mo. Intermetallics 2003, 11, 71–82.

36 Kubaschewski, O. Iron-Binary Phase Diagrams; Springer: Berlin, Germany, 1982.

37 Palm, M. Concepts derived from phase diagram studies for the strengthening of Fe-Al-based alloys. Intermetallics2005, 13, 1286–1295.

38 Morris, D.; Liu, C.; George, E. Pinning of dislocations and the origin of the stress anomaly in FeAl alloys. Intermetallics1999, 7, 1059–1068.

39 Morris, D.G.; Munoz-Morris, M. The stress anomaly in FeAl-Fe3Al alloys. Intermetallics 2005, 13, 1269–1274.

40 Friák, M.; Deges, J.; Krein, R.; Frommeyer, G.; Neugebauer, J. Combined ab initio and experimental study of structural and elastic properties of Fe3Al-based ternaries. Intermetallics 2010, 18, 1310–1315.

41 Palm, M.; Inden, G.; Thomas, N. The Fe-Al-Ti system. J. Phase Equilib. 1995, 16, 209–222.

42 Ducher, R.; Stein, F.; Viguier, B.; Palm, M.; Lacaze, J. A re-examination of the liquidus surface of the Al-Fe-Ti system.Z. Metallkd. 2003, 94, 396–410.

43 Calvert, L.; Villars, P. Pearson's Handbook of Crystallographic Data for Intermetallic Phases; ASM International: Materials Park, OH, USA, 1991.

44 Shu, X.L.; Hu, W.Y.; Xiao, H.N.; Deng, H.Q.; Zhang, B.W. Vacancies and antisites in B2 FeAl and D03 Fe3Al with a modified analytic eam model. J. Mater. Sci. Technol. 2001, 17, 601–604.

45 Leamy, H.; Gibson, E.; Kayser, F. The elastic stiffness coefficients of iron-aluminum alloys-I experimental results and thermodynamic analysis. Acta Metall. Mater. 1967, 15, 1827–1838.

46 Vailhe, C.; Farkas, D. Shear faults and dislocation core structure simulations in B2 FeAl. Acta Mater. 1997, 45, 4463–4473.

47 Stein, F. Summer School on Iron Aluminides Part 1: The binary Fe-Al system. In Proceedings of the 5th Discussion Meeting on the Development of Innovative Iron Aluminum Alloys in Prague, Prague, Czech Republic, 21–24 September 2009.

48 Bradley, A.; Jay, A. The formation of superlattices in alloys of iron and aluminium. Proc. R. Soc. Lond. Ser. A 1932, 210–232.

49 Lawley, A.; Cahn, R. A high temperature X-ray study of ordering in iron-aluminium alloys. J. Phys. Chem. Solids 1961,20, 204 221.

50 Fu, C.L. Origin of ordering in B2-type transition-metal aluminides: Comparative study of the defect properties of PdAl, NiAl, and FeAl. Physical Review B 1995, 52, 3151–3158.

51 Herrmann, J.; Inden, G.; Sauthoff, G. Deformation behaviour of iron-rich iron-aluminum alloys at low temperatures.Acta Mater. 2003, 51, 2847–2857.

52 Herrmann, J.; Inden, G.; Sauthoff, G. Deformation behaviour of iron-rich iron-aluminium alloys at high temperatures.Acta Mater. 2003, 51, 3233–3242.

53 Hasemann, G.; Schneibel, J.H.; George, E.P. Dependence of the yield stress of Fe3Al on heat treatment. Intermetallics2012, 21, 56–61.

54 Hasemann, G.; Schneibel, J.H.; Krüger, M.; George, E.P. Vacancy strengthening in Fe3Al iron aluminides. Intermetallics2014, 54, 95–103.

55 Neumann, J.; Chang, Y.; Lee, C. Thermodynamics of intermetallic phases with the triple-defect B2 structure. Acta Metall. Mater. 1976, 24, 593–604.

56 Foiles, S.; Daw, M. Application of the embedded atom method to Ni3Al. J. Mater. Res. 1987, 2, 5–15.

57 Fu, C.; Ye, Y.-Y.; Yoo, M.; Ho, K. Equilibrium point defects in intermetallics with the B2 structure: NiAl and FeAl.Phys. Rev. B 1993, 48, 6712.

58 Stein, F.; Palm, M. Re-determination of transition temperatures in the Fe-Al system by differential thermal analysis.Int J. Mater. Res. 2007, 98, 580–588.

59 Schaefer, H.-E.; Frenner, K.; Würschum, R. High-temperature atomic defect properties and diffusion processes in intermetallic compounds. Intermetallics 1999, 7, 277–287.

60 Hehenkamp, T.; Scholz, P.; Köhler, B.; Kerl, R. Vacancy formation and diffusion in FeAl-alloys. Defect Diffus. Forum2001, 194–199, 389–396.

61 Würschum, R.; Grupp, C.; Schaefer, H.-E. Simultaneous study of vacancy formation and migration at high temperatures in B2-type Fe aluminides. Phys. Rev. Lett. 1995, 75, 97. [PubMed]

62 Schneibel, J. Strengthening of iron aluminides by vacancies and/or nickel. Mater. Sci. Eng. A 1998, 258, 181–186.

63 Mayer, J.; Meyer, B.; Oehrens, J.S.; Bester, G.; Börnsen, N.; Fähnle, M. Effective formation energies of atomic defects in D03 Fe3Al: An ab-initio study. Intermetallics 1997, 5, 597–600.

64 Fähnle, M.; Mayer, J.; Meyer, B. Theory of atomic defects and diffusion in ordered compounds, and application to B2-FeAl. Intermetallics 1999, 7, 315–323.

65 Besson, R.; Morillo, J. Development of a semiempirical n-body noncentral potential for Fe-Al alloys. Phys. Rev. B 1997,55, 193–204.

66 Bakker, H.; Modder, I.; Kuin, M. Extension of miedema's semiempirical model to estimates of the formation enthalpies of point defects in intermetallic compounds with the B2 structure. Intermetallics 1997, 5, 535–546.

67 Yasuda, H.Y.; Umakoshi, Y. Pseudoelastic behaviour of Fe3Al single crystals with D03 structure. Intermetallics 2010,18, 1273–1278.

68 Yasuda, H.; Nakajima, T.; Nakano, K.; Yamaoka, K.; Ueda, M.; Umakoshi, Y. Effect of Al concentration on pseudoelasticity in Fe3Al single crystals. Acta Mater. 2005, 53, 5343–5351.

69 Yasuda, H.Y.; Nakajima, T.; Umakoshi, Y. Temperature dependence of pseudoelasticity in Fe3Al single crystals.Intermetallics 2007, 15, 819–823.

70 Wolff, J.; Franz, M.; Hehenkamp, T. Defect analysis with positron annihilation-Applications to Fe aluminides.Microchim. Acta 1997, 125, 263–268.

71 Saburi, T.; Yamauchi, I.; Nenno, S. Electron microscope observation of dislocations and antiphase boundaries in iron-aluminum alloys. J. Phys. Soc. Jpn. 1972, 32, 694–701.

72 Crawford, R.C.; Ray, I.L.F.; Cockayne, D.J. Weak-beam technique applied to superlattice dislocations in iron-aluminum alloys. 2. Fourfold dissociation in D03-type order. Philos. Mag. 1973, 27, 1–7.

73 Crawford, R.C.; Ray, I.L.F. Antiphase boundary energies in iron-aluminum alloys. Philos. Mag. 1977, 35, 549–565.

74 Král, F.; Schwander, P.; Kostorz, G. Superdislocations and antiphase boundary energies in deformed Fe3Al single crystals with chromium. Acta Mater. 1997, 45, 675–682.

75 Morris, D.; Dadras, M.; Morris, M. The influence of cr addition on the ordered microstructure and deformation and fracture behaviour of a Fe 28% Al intermetallic. Acta Metall. Mater. 1993, 41, 97–111.

76 Brinck, A.; Engelke, C.; Neuhäuser, H.; Molénat, G.; Rösner, H.; Langmaack, E.; Nembach, E. Dislocation processes in Fe3Al investigated by transmission electron, scanning force and optical microscopy. Mater. Sci. Eng. A 1998, 258, 32–36.

77 Yoshimi, K.; Hanada, S.; Yoo, M. Yielding and plastic flow behavior of B2-type Fe-39.5 mol.% Al single crystals in compression. Acta Metall. Mater. 1995, 43, 4141–4151.

78 Schröer, W.; Hartig, C.; Mecking, H. Plasticity of D03-ordered Fe-Al and Fe-Al-Si single-crystals. Z. Metallkd. 1993, 84, 294–300.

79 Yoo, M.; Koeppe, M.; Hartig, C.; Mecking, H.; Hermann, W.; Sockel, H.-G. Effect of temperature on elastic constants and dislocation properties of Fe 30% Al single crystals. Acta Mater. 1997, 45, 4323–4332.

80 Yoo, M.; Horton, J.; Liu, C. Micromechanisms of yield and flow in ordered intermetallic alloys. Acta Metall. Mater.1988, 36, 2935–2946.

81 Koster, W.; Godecke, T. Physical measurements on iron-aluminium alloys between 10 and 50 at.-% Al. Iv.—The modulus of elasticity of the alloys. Z. Metallkd. 1982, 73, 111–114.

82 Palm, M. The Al-Cr-Fe system–phases and phase equilibria in the Al-rich corner. J. Alloys Compd. 1997, 252, 192–200.

83 Alonso, P.R.; Gargano, P.H.; Bozzano, P.B.; Ramírez-Caballero, G.E.; Balbuena, P.B.; Rubiolo, G.H. Combined ab initio and experimental study of A2 + L21 coherent equilibria in the Fe-Al-X (X = Ti, Nb, V) systems. Intermetallics 2011, 19, 1157–1167.

84 Stein, F.; Sauthoff, G.; Palm, M. Phases and phase equilibria in the Fe-Al-Zr system: Dedicated to professor dr. Peter neumann on the occasion of his 65th birthday. Z. Metallkd. 2004, 95, 469–485.

85 Fu, C.L.; Zou, J. Site preference of ternary alloying additions in FeAl and NiAl by first-principles calculations. Acta. Mater 1996, 44, 1471–1478.

86 Zuqing, S.; Wangyue, Y.; Lizhen, S.; Yuanding, H.; Baisheng, Z.; Jilian, Y. Neutron diffraction study on site occupation of substitutional elements at sub lattices in Fe3Al intermetallics. Mater. Sci. Eng.: A 1998, 258, 69–74.

87 Nishino, Y.; Asano, S.; Ogawa, T. Phase stability and mechanical properties of Fe3Al with addition of transition elements. Mater. Sci. Eng.: A 1997, 234, 271–274.

88 Nishino, Y.; Kumada, C.; Asano, S. Phase stability of Fe3Al with addition of 3d transition elements. Scr. Mater. 1997,36, 461–466.

89 Krein, R.; Schneider, A.; Sauthoff, G.; Frommeyer, G. Microstructure and mechanical properties of Fe3Al-based alloys with strengthening boride precipitates. Intermetallics 2007, 15, 1172–1182.

90 Connetable, D.; Lacaze, J.; Maugis, P.; Sundman, B. A calphad assessment of Al C Fe system with the κ carbide modelled as an ordered form of the fcc phase. Calphad 2008, 32, 361–370.

91 Schneider, A.; Falat, L.; Sauthoff, G.; Frommeyer, G. Constitution and microstructures of Fe-Al-M-C (M = Ti, V, Nb, Ta) alloys with carbides and laves phase. Intermetallics 2003, 11, 443–450.

92 Kobayashi, S.; Zaefferer, S.; Schneider, A. Carbide Precipitation in an Fe3Al-Cr-Mo-C Alloy. In Proceedings of an International Conference on Solid-Solid Phase Transformations in Inorganic

Materials 2005, Vol. 1: Diffusional transformations; Wiley: New York, NY, USA, 2005.

93 Doucakis, T.; Kumar, K. Formation and stability of refractory metal diborides in an Fe3Al matrix. Intermetallics 1999, 7, 765–777.

94 Morris, M.; Morris, D. Dispersoid additions and their effect on high temperature deformation of FeAl. Acta Metall. Mater. 1990, 38, 551–559.

95 Palm, M. Fe-Al materials for structural applications at high temperatures: Current research at mpie. Int. J. Mater. Res.2009, 100, 277–287.

96 Risanti, D.; Deges, J.; Falat, L.; Kobayashi, S.; Konrad, J.; Palm, M.; Pöter, B.; Schneider, A.; Stallybrass, C.; Stein, F. Dependence of the brittle-to-ductile transition temperature (BDTT) on the Al content of Fe-Al alloys. Intermetallics2005, 13, 1337–1342.

97 Anthony, L.; Fultz, B. Effects of early transition metal solutes on the D03-B2 critical temperature of Fe3Al. Acta Metall. Mater. 1995, 43, 3885–3891.

98 Fleischer, R.L. Substitutional solutes in AlRu-I. Effects of solute on moduli, lattice parameters and vacancy production. Acta Metall. Mater. 1993, 41, 863–869.

99 Fleischer, R.L. Substitutional solutes in AlRu-II. Hardening and correlations with defect structure. Acta Metall. Mater.1993, 41, 1197–1205.

100 Schneibel, J.; Specht, E.; Simpson, W. Solid solution strengthening in ternary B2 iron aluminides containing 3d transition elements. Intermetallics 1996, 4, 581–583.

101 Zamanzade, M.; Barnoush, A. Effect of chromium on the electrochemical properties of iron aluminide intermetallics.Corros. Sci. 2014, 78, 223–232.

102 Rao, V.S. A review of the electrochemical corrosion behaviour of iron aluminides. Electrochim. Acta 2004, 49, 4533–4542.

103 Balasubramaniam, R. On the role of chromium in minimizing room temperature hydrogen embrittlement in iron aluminides. Scr. Mater. 1996, 34, 127–133.

104 Hakiki, N. Comparative study of structural and semiconducting properties of passive films and thermally grown oxides on AISI 304 stainless steel. Corros. Sci. 2011, 53, 2688–2699.

105 Rose, J.H.; Smith, J.R.; Ferrante, J. Universal features of bonding in metals. Phys. Rev. B 1983, 28, 1835.

106 Rose, J.H.; Smith, J.R.; Guinea, F.; Ferrante, J. Universal features of the equation of state of metals. Phys. Rev. B 1984,29, 2963.

107 Morris, D.G.; Dadras, M.; Morris, M. The influence of chromium additions on order and ductility in Fe3Al intermetallic. J. Phys. IV 1993, 3, C7-429–C427-434.

108 Zamanzade, M.; Vehoff, H.; Barnoush, A. Effect of chromium on elastic and plastic deformation of Fe3Al intermetallics. Intermetallics 2013, 41, 28–34.

109 Zamanzade, M.; Rafael-Velayarce, J.; Torrents-Abad, O.; Motz, C.; Barnoush, A. Mechanical behavior of iron aluminides: A comparison of nanoindentation, compression and bending of micropillars. Mater. Sci. Eng. A 2015.

110 Barnoush, A.; Zamanzade, M. Effect of substitutional solid solution on dislocation nucleation in Fe3Al intermetallic alloys. Philos. Mag. 2012, 92, 3257–3268.

111 Motz, C.; Schoberl, T.; Pippan, R. Mechanical properties of micro-sized copper bending beams machined by the focused ion beam technique. Acta Mater. 2005, 53, 4269–4279.

112 Ellinger, C.S. In situ oxidation study of flat and stepped binary alloy surfaces. Ph.D. Thesis, Institut für Theoretische und Angewandte Physik &Max-Planck-Institut für Intelligente Systeme, Stuttgart, Germany, 2010.

113 Hammer, L.; Meier, W.; Blum, V.; Heinz, K. Equilibration processes in surfaces of the binary alloy Fe-Al. J. Phys. Condens. Matter 2002, 14.

114 Baddorf, A.; Chandavarkar, S. Identification of an incommensurate FeAl2 overlayer on FeAl (110) using X-ray diffraction and reflectivity. Phys. B: Condens. Matter 1996, 221, 141–144.

115 Voges, D.; Taglauer, E.; Dosch, H.; Peisl, J. Surface segregation on Fe3Al (110) near the order-disorder transition temperature. Surf. Sci. 1992, 269, 1142–1146.

116 Dosch, H.; Mailänder, L.; Johnson, R.; Peisl, J. Critical phenomena at the Fe3Al (110) surface: A glancing angle X-ray scattering study. Surf. Sci. 1992, 279, 367–379.

117 Hutchings, R.; Loretto, M. Compositional dependence of oxidation rates of NiAl and CoAl. Metal Sci. 1978, 12, 503–510.

118 Pöter, B.; Stein, F.; Wirth, R.; Spiegel, M. Early stages of protective oxide layer growth on binary iron aluminides. Z. Phys. Chem. 2005, 219, 1489–1503.

119 Velon, A.; Olefjord, I. Oxidation behavior of Ni3Al and Fe3Al: I. Early stage of oxide growth. Oxid. Metals 2001, 56, 425–452.

120 Velon, A.; Yi, D.-Q. Influence of Cr on the oxidation of Fe3Al and Ni3Al at 500 °C. Oxid. Metals 2002, 57, 13–31.

121 Lee, W.; Lin, R. Oxidation, sulfidation and hot corrosion of intermetallic compound Fe3Al at 605 °C and 800 °C. Mater. Chem. Phys. 1999, 58, 231–242.

122 Vogel, D.; Hotař, A.; Vogel, A.; Palm, M.; Renner, F.U. Corrosion behaviour of Fe-Al(-Ti) alloys in steam. Intermetallics2010, 18, 1375–1378.

123 Pint, B.A.; Wright, I.G. The Oxidation Behavior of Fe-Al Alloys. Mater. Sci. Forum 2004, 461–464, 799–806.

124 Janda, D. Mechanical Properties and Oxidation Behavior of Micro-Alloyed Iron Aluminides. Ph.D. Thesis, Karlsruher Institut für Technologie (KIT), Karlsruhe, Germany, 18 February 2015.

125 Grabke, H. Oxidation of NiAl and FeAl. Intermetallics 1999, 7, 1153–1158.

126 Kim, I.; Cho, W.D.; Kim, H.J. High-temperature oxidation of Fe3Al containing yttrium. J. Mater. Sci. 2000, 35, 4695–4703.

127 Frangini, S.; Giorgi, R.; Lascovich, J.; Mignone, A. XPS study of passive films formed on an iron-aluminium intermetallic compound in acid solution. Surf. Interface Anal. 1994, 21, 435–441.

128 McCafferty, E. Thermodynamics of corrosion: Pourbaix diagrams. In Introduction to Corrosion Science; Springer: Berlin, Germany, 2010; pp. 95–117.

129 De Cristofaro, N.; Frangini, S.; Mignone, A. Passivity and passivity breakdown on a β-FeAl intermetallic compound in sulphate and chloride containing solutions. Corros. Sci. 1996, 38, 307–315.

130 Frangini, S.; De Cristofaro, N.; Lascovich, J.; Mignone, A. On the passivation characteristics of a β-FeAl intermetallic compound in sulphate solutions. Corros. Sci. 1993, 35, 153–159.

131 Moshier, W.; Davis, G.; Ahearn, J. The corrosion and passivity of aluminum exposed to dilute sodium sulfate solutions. Corros. Sci. 1987, 27, 785–801.

132 George, F.O. Chromium-Free Conversion Coating of Aluminium-Copper Alloys. Ph.D. Thesis, The University of Manchester, Manchester, UK, 4 February 2011.

133 Vargel, C. Corrosion of Aluminium; Elsevier: Amsterdam, The Nietherlands, 2004.

134 García-Alonso, M.C.; López, M.F.; Escudero, M.L.; González-Carrasco, J.L.; Morris, D.G. Corrosion behaviour of an Fe3Al-type intermetallic in a chloride containing solution. Intermetallics 1999, 7, 185–191.

135 Rosalbino, F.; Carlini, R.; Parodi, R.; Zanicchi, G.; Scavino, G. Investigation of passivity and its breakdown on Fe3Al-Si and Fe3Al-Ge intermetallics in chloride-containing solution. Corros. Sci. 2014, 85, 394–400.

136 Johnson, W.H. On some remarkable changes produced in iron and steel by the action of hydrogen and acids. Proc. R. Soc. Lond. 1874, 23, 168–179.

137 Lynch, S. Comments on "a unified model of environment-assisted cracking". Scr. Mater. 2009, 61, 331–334.

138 Gangloff, R.P.; Somerday, B.P. Gaseous Hydrogen Embrittlement of Materials in Energy Technologies: Mechanisms, Modelling and Future Developments; Elsevier: Amsterdam, The Nietherlands, 2012; Volume 2.

139 McKamey, C.; Liu, C. Chromium addition and environmental embrittlement in Fe3Al. Scr. Metall. Mater. 1990, 24, 2119–2122.

140 Stroe, M.E. Hydrogen Embrittlement of Ferrous Materials. Ph.D. Thesis, Universite Libre De Bruxelles, Brussel, Belgium, 31 March 2006.

141 Fu, C.; Painter, G. First principles investigation of hydrogen embrittlement in FeAl. J. Mater. Res. 1991, 6, 719–723.

142 Johnson, D.F.; Carter, E.A. First-principles assessment of hydrogen absorption into FeAl and Fe3Si: Towards prevention of steel embrittlement. Acta Mater. 2010, 58, 638–648.

143 Hirth, J.P. Effects of hydrogen on the properties of iron and steel. Metall. Trans. A 1980, 11, 861–890.

144 Cheng, X.; Wan, X. Hydrogen diffusivity in a Fe3Al-based alloy. Scr. Mater. 1998, 38, 1505–1509.

145 Kupka, M.; Stępień, K. Hydrogen permeation in Fe-40at.% Al alloy at different temperatures. Corros. Sci. 2009, 51, 699–702.

146 Banerjee, P.; Balasubramaniam, R. Hydrogen diffusivity in iron aluminides determined by subscale microhardness profiling. Scr. Mater. 1998, 39, 1215–1219.

147 Yang, Y.; Hanada, S. Absorption and desorption of hydrogen in Fe-40Al intermetallic. Scr. Metall. Mater. 1995, 32, 1719–1724.

148 Luu, W.; Wu, J. Hydrogen transport and environmental embrittlement effects in iron aluminides. J. Mater. Sci. 2000,35, 4121–4127.

149 Stępień, K.; Kupka, M. Diffusivity of hydrogen in B2 iron aluminides. Scr. Mater. 2006, 55, 585–588.

150 Prakash, U.; Parvathavarthini, N.; Dayal, R. Effect of composition on hydrogen permeation in Fe-Al alloys.Intermetallics 2007, 15, 17–19.

151 Kimizuka, H.; Mori, H.; Ogata, S. Effect of temperature on fast hydrogen diffusion in iron: A path-integral quantum dynamics approach. Phys. Rev. B 2011, 83, 094110.

152 Bockris, J.M.; Fullenwider, M. The electro-permeation of hydrogen into metals. Electrochim. Acta 1970, 15, 47–60.

153 Song, R.H.; Pyun, S.; Oriani, R. Hydrogen permeation through the passivation film on iron by time-lag method. J. Electrochem. Soc. 1990, 137, 1703–1706.

154 Ruetschi, P.; Giovanoli, R. Cation vacancies in MnO2 and their influence on electrochemical reactivity. J. Electrochem. Soc. 1988, 135, 2663–2669.

155 Song, R.-H.; Pyun, S.-I.; Oriani, R. The hydrogen permeation through passivating film on iron by modulation method.Electrochim. Acta 1991, 36, 825–831.

156 Zhang, G.; Wang, X.; Yang, F.; Shi, Y.; Song, J.; Lai, X. Energetics and diffusion of hydrogen in hydrogen permeation barrier of α-Al2O3/ FeAl with two different interfaces. Int. J. Hydrog. Energy 2013, 38, 7550–7560.

157 Hollenberg, G.; Simonen, E.; Kalinin, G.; Terlain, A. Tritium/hydrogen barrier development. Fusion Eng. Des. 1995, 28, 190–208.

158 Barnoush, A.; Vehoff, H. Hydrogen embrittlement of aluminum in aqueous environments examined by in situelectrochemical nanoindentation. Scr. Mater. 2008, 58, 747–750.

159 Pundt, A.; Kirchheim, R. Hydrogen in metals: Microstructural aspects. Annu. Rev. Mater. Res. 2006, 36, 555–608.

160 Myers, S.M.; Baskes, M.; Birnbaum, H.; Corbett, J.W.; De Leo, G.; Estreicher, S.; Haller, E.E.; Jena, P.; Johnson, N.M.; Kirchheim, R. Hydrogen interactions with defects in crystalline solids. Rev. Mod. Phys. 1992, 64, 559–617.

161 Ryu, J.H. Hydrogen Embrittlement in Trip and Twip Steels. Ph.D. Thesis, Pohang University of Science and Technology, Pohang, South Korea, 29 May 2012.

162 Myers, S.; Besenbacher, F.; No, J. Immobilization mechanisms for ion-implanted deuterium in aluminum. J. Appl. Phys. 1985, 58, 1841–1850.

163 Ziegler, R.; Schaefer, H. Vacancies and interstitials in metals and alloys. In Proceedings of the 1987 International Conference on Vacancies and Interstititials in Metals and Alloys, Berlin, Germany, 14–19 September 1987.

164 Linderoth, S.; Rajainmäki, H.; Nielsen, B.; Hansen, H.; Nieminen, R.M.; Petersen, K. Hydrogen in Vacancies in Mo and Ni. Mater. Sci. Forum 1987, 15–18, 751–756.

165 Wittmann, M.; Wu, D.; Baker, I.; George, E.; Heatherly, L. The role of edge and screw dislocations on hydrogen embrittlement of Fe-40Al. Mater. Sci. Eng.: A 2001, 319, 352–355.

166 Tien, J.; Thompson, A.W.; Bernstein, I.; Richards, R.J. Hydrogen transport by dislocations. Metall. Trans. A 1976, 7, 821–829.

167 Louthan, M.; Caskey, G.; Donovan, J.; Rawl, D. Hydrogen embrittlement of metals. Mater. Sci. Eng. 1972, 10, 357–368.

168 Donovan, J.A. Accelerated evolution of hydrogen from metals during plastic deformation. Metall. Trans. A 1976, 7, 1677–1683.

169 Nathal, M.; Liu, C. Intrinsic ductility of FeAl single crystals. Intermetallics 1995, 3, 77–81.

170 Lynch, R.; Gee, K.; Heldt, L. Environmental embrittlement of single crystal and thermomechanically processed B2-ordered iron aluminides. Scr. Metall. Mater. 1994, 30, 945–950.

171 Saka, H.; Nishizaki, T. Roles of edge and screw dislocations in the environmental embrittlement of an Fe [sbnd] Al intermetallic compound. Philos. Mag. A 1996, 73, 1173–1179.

172 Yang, Y.; Baker, I.; George, E. Effect of vacancies on the tensile properties of Fe-40Al single crystals in air and vacuum.Mater. Charact. 1999, 42, 161–167.

173 Munroe, P.; Baker, I. Observation of <001> dislocations and a mechanism for transgranular fracture on {001} in FeAl.Acta Metall. Mater. 1991, 39, 1011–1017.

174 Li, J.; Liu, C. Crack nucleation in hydrogen embrittlemet. Scr. Metall. Mater. 1992, 27, 1701–1705.

CHAPTER 10

Effect of Heat Treatment Process on Microstructure and Fatigue Behavior of a Nickel-Base Superalloy

Peng Zhang [1], Qiang Zhu [1], Gang Chen [1], Heyong Qin [2] and Chuanjie Wang [1,*]

[1] School of Materials Science and Engineering, Harbin Institute of Technology at Weihai, Weihai 264209, China

[2] Central Iron & Steel Research Institute, Beijing 100081, China

ABSTRACT

The study of fatigue behaviors for nickel-base superalloys is very significant because fatigue damage results in serious consequences. In this paper, two kinds of heat treatment procedures (Pro.I and Pro.II) were taken to investigate the effect of heat treatment on microstructures and fatigue behaviors of a nickel-base superalloy. Fatigue behaviors were studied through total strain controlled mode at 650 °C. Manson-Coffin relationship and three-parameter power function were used to predict fatigue life. A good link between the cyclic/fatigue behavior and microscopic studies was established. The cyclic deformation mechanism and fatigue mechanism were discussed. The results show that the fatigue resistance significantly drops with the increase of total strain amplitudes. Manson-Coffin relationship can well predict the fatigue life for total strain amplitude from 0.5% to 0.8%. The fatigue resistance is related with heat treatment procedures.

The fatigue resistance performance of Pro.I is better than that of Pro.II. The cyclic stress response behaviors are closely related to the changes of the strain amplitudes. The peak stress of the alloy gradually increases with the increase of total strain amplitudes. The main fracture mechanism is inhomogeneous deformation and the different interactions between dislocations and γ' precipitates.

KEYWORDS

heat treatment; microstructure; fatigue behavior; fatigue life model; fracture mechanism

1. INTRODUCTION

Nickel-base superalloys have become critical materials for high temperature components of gas-turbine engines due to high temperature strength, oxidation resistance, excellent creep resistance, excellent fatigue resistance at elevated temperature [1,2,3]. Turbine disk, an important high temperature component of gas-turbine engines, withstands effects of high temperatures and alternating loading in actual service conditions. In this case, strain-controlled high temperature low cycle fatigue (LCF) damage occurs for turbine disk. Fatigue damage seriously affects the persistent and normal use of high temperature components. In addition, fatigue damage reduces their service lives and leads to catastrophic consequences [4,5]. Thus, the study of fatigue behaviors for nickel-base superalloys is necessary and significant. Numerous studies have been carried out about the effects of cyclic frequency, inclusions, microstructure, strain range, mean stress, pre-deformation, environment and testing temperature on fatigue behaviors of nickel-base superalloys [6,7,8,9,10].

The alloy studied is a precipitation-hardened high nickel superalloy, which is used extensively in the gas-turbine engines. The effect of high nickel is to improve the plasticity and ductility of the superalloy. In order to get a perfect overall performance of turbine disks, especially excellent fatigue resistance, heat treatment process is an important factor in addition to its composition and appropriate manufacturing process. The appropriate heat treatment process can give full play to the potential of alloys and ensure its safe and reliable work. The effects of heat treatment process on microstructure and fatigue behavior of nickel-base superalloys have been less investigated so far. Du *et al.* [11] investigated the effects of the solution and aging treatment on the microstructure of GH4169 superalloy. James and Mills [12] studied the fatigue-crack growth behavior of Alloy 718 at different test temperatures. Electron fractographic examination revealed that operative crack growth mechanisms were dependent on heat treatment, heat-to-heat variations, testing temperature and stress-intensity level of crack tip. In this paper, the effects of heat treatment process on microstructure and fatigue behavior of the alloy were simultaneously investigated. Two kinds of heat treatment procedures were taken for the nickel-base superalloy. Then fatigue tests were conducted for the specimens selected from the heat treatment procedures. It is of great significance to study the effect of different solution treatments on microstructures and fatigue behaviors of the superalloy. As for gas-turbine engine turbine disks, the actual service temperature is about 650 °C. LCF behaviors of the alloy were studied at 650 °C under condition of total strain-controlled mode. The effect of strain amplitudes on cycle mechanical response was studied. Moreover, the microstructures of different heat treatment procedures after fatigue test were investigated. A good link between the cyclic/fatigue behavior and microscopic studies was established. It would be much meaningful to improve the production of nickel-base superalloys and fatigue life of turbine disks through the study of LCF behaviors for the nickel-base superalloy by different heat treatment procedures.

2. EXPERIMENTAL SECTION

The nickel-base superalloy was smelted by vacuum induction furnace and re-melted by vacuum consumable electrode melting furnace. In order to obtain uniformity of the grain size, the nickel base superalloy was forged at 1100 °C. The chemical composition (wt %) of the alloy is as follows: C 0.042, Cr 14.50, Mo 3.18, Al 1.70, Ti 2.68, Fe < 0.10, Nb 2.02, and Ni the rest.

2.1. Heat Treatment Procedure

In order to improve the mechanical properties of the alloy and eliminate residual stress, especially to investigate the effect of heat treatment process on LCF behavior, two kinds of heat treatment procedures were taken, which are as follows:

- ProcedureI (Pro.I): 1050 °C/8 h, AC + 1000 °C/4 h, AC + 775 °C/16 h, AC + 700 °C/16 h, AC
- ProcedureII (Pro.II): 1100 °C/8 h, AC + 1000 °C/4 h, AC + 775 °C/16 h, AC + 700 °C/16 h, AC

Where, AC means air cooling. Heat treatment procedures are four stages. The first stage is solution treatment. For example, the stage "1050 °C/8 h, AC" means that the specimens were heated from room temperature to 1050 °C and heat preservation time was 8 h at a constant temperature of 1050 °C. Then the specimens were air-cooled to room temperature. The second stage (1000 °C/4 h, AC) is intermediate heat treatment and the last two stages (775 °C/16 h, AC + 700 °C/16 h, AC) are aging treatment.

2.2. Fatigue Tests

The standard of fatigue tests is referred to ISO 1099:2006 (Metallic materials-Fatigue testing-Axial-force-controlled method) [13]. The fatigue specimens selected from heat-treated specimens were machined into 6.350 mm in diameter and 16 mm in gauge length. LCF tests were carried out on MTS 810 fatigue testing machine (MTS Systems Corporation, Eden Prairie, MN, USA) in air. Experimental program was total axial strain-controlled from 0.3% to 0.8% at constant temperature of 650 °C. The strain ratio was $R = -1$ and the loading frequency was 0.50 Hz. The fatigue specimens were mechanical grinded, polished and chemical etched so as to observe grain morphology of the nickel-base superalloy by Olympus microscope. The precipitates of the nickel-base superalloy after heat treatment were observed by Quanta 200FEG scanning electron microscopy (SEM, FEI Company, Hillsboro, OR, USA). In order to investigate the fracture mechanism of the alloy, the fracture morphology of fatigue failure specimens was observed by SEM and the dislocation characteristics were observed by Tecnai G2 F30 transmission electron microscopy (TEM, FEI Company, Hillsboro, OR, USA). Specimens for TEM were obtained from thin slices (500 μm in thickness) at a distance of 1 mm away from the fracture surfaces of the failed specimens. Thin slices were ground to about 50 μm, and then prepared for twin-jet electrolytic thinning.

3. CYCLIC STRESS RESPONSE BEHAVIOR AND DEFORMATION MECHANISM

3.1. Grain Morphology and Precipitates

The purpose of solution treatment is to dissolve the original reinforcing phases, to dissolve the irrational distributed secondary carbide and boride phases on grain boundary, to obtain the desired grain size and to eliminate residual stresses. The purpose of the aging treatment is to precipitate evenly fine dispersive distributed γ′ phases, at the same time to precipitate the complex carbide and boride phases on grain boundary. Turbine disks will possess a perfect overall performance after heat treatment. Figure 1 shows the grain morphology of the alloy. It can be observed that the average size of the original grain is the smallest and average grain size of Pro.I is much smaller than that of Pro.II. In addition, many twins exist in the original grain and some twins still exist after heat treatment. The grain sizes are measured by intersection point method. The average sizes of original grain, Pro.I and Pro.II are 31.4 μm, 70.7 μm and 144.3 μm, respectively. The precipitates of the nickel-base superalloy after heat treatment are shown in Figure 2. The incipient melting temperature of γ′ phase for the alloy is 1050 °C and fully melting temperature of γ′ phase is 1080 °C. As for Pro.I, solution temperature is lower than fully melting temperature of γ′ phase. As for Pro.II, the solution temperature is higher than fully melting temperature of γ′ phase. It can be observed from Figure 2 that the distribution of γ′ phases for Pro.II is more uniform than that for Pro.I. It shows that undissolved γ' phases in the solution treatment process hinder grain growth. The purpose of the intermediate heat treatment is to make small γ' phases formed during cooling after solution treatment grow up as ideal large γ' phases and to re-precipitate small γ' phases. There are some large carbide particles at the grain boundaries.

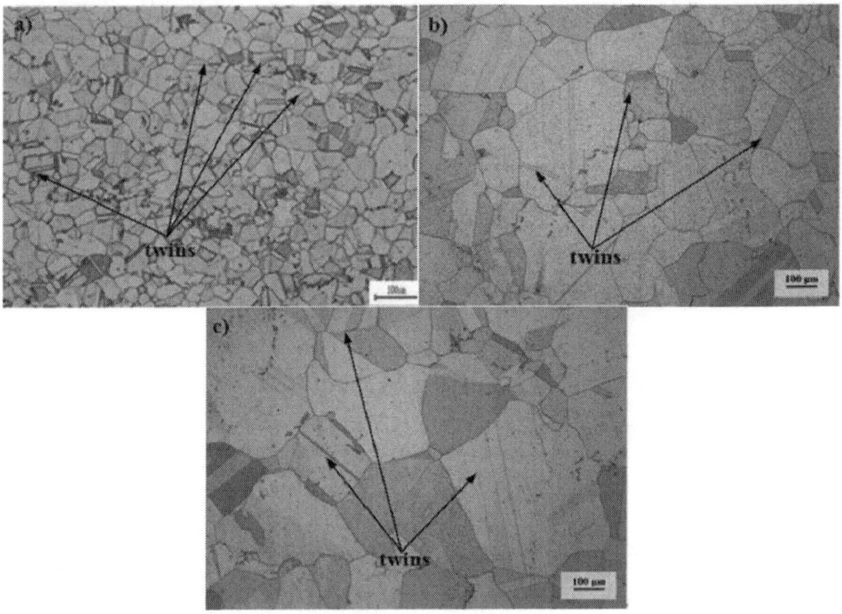

Figure 1. The grain morphology of the nickel-base superalloy. (**a**) the original grain; (**b**) Pro.I; (**c**) Pro.II.

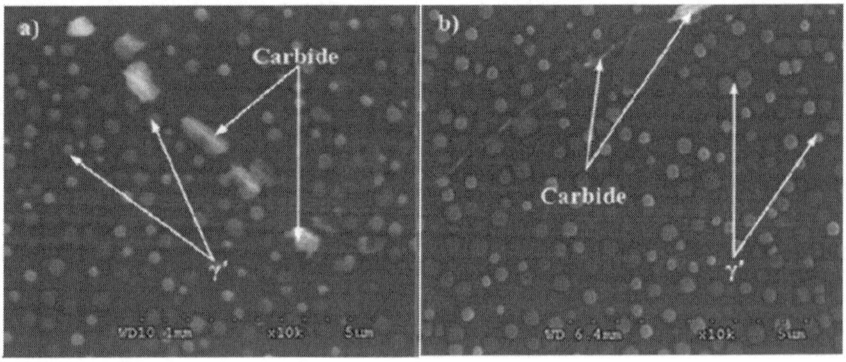

Figure 2. The precipitates of the nickel-base superalloy after heat treatment. (**a**) Pro.I; (**b**) Pro.II.

3.2. Fatigue Property and Fatigue Life Model

The fatigue test results under different heat treatment procedures are shown in Table 1, where, $\Delta\varepsilon_t/2$ is the total strain, $\Delta\varepsilon_e/2$ is the elastic strain, $\Delta\varepsilon_p/2$ is the peak plastic strain, and $2N_f$ is the reverse number of fatigue cycles. It can be found that fatigue resistance significantly drops with the increase of strain amplitudes under the condition of the same heat treatment procedure. It can be observed that the fatigue life of Pro.I is higher than that of Pro.II at the same total strain amplitude, which indicates that the heat treatments play an important role in fatigue behavior and the fatigue resistance of Pro.I is better than that of Pro.II.

For strain-controlled LCF, total strain amplitude includes elastic strain and plastic strain. The Manson-Coffin relationship can be used to describe the strain-controlled LCF behavior. According to Manson-Coffin relationship and Basqin relationship, the relationship of the elastic, plastic and total strain ranges is expressed as follow [14]:

$$\frac{\Delta\varepsilon_t}{2} = \frac{\Delta\varepsilon_e}{2} + \frac{\Delta\varepsilon_p}{2} = \frac{\sigma'_f}{E}(2N_f)^b + \varepsilon'_f(2N_f)^c \qquad (1)$$

where σ'_f is the fatigue strength coefficient; ε'_f is the fatigue ductility coefficient; b is the fatigue strength exponent, c is the fatigue ductility exponent; E is Young's modulus.

Table 1. Fatigue test results under different heat treatment procedures.

Heat Treatment	$\Delta\varepsilon_t/2$ (%)	$\Delta\varepsilon_e/2$ (%)	$\Delta\varepsilon_p/2$ (%)	$2N_f$
Pro.I	0.3	0.297	0.003	68,468
	0.4	0.378	0.022	8094
	0.5	0.413	0.087	1746
	0.6	0.473	0.127	768
	0.7	0.489	0.211	282
	0.8	0.51	0.29	174
Pro.II	0.3	0.297	0.003	52,032
	0.4	0.368	0.032	4528
	0.5	0.414	0.086	1622
	0.6	0.466	0.134	648
	0.7	0.491	0.209	242
	0.8	0.508	0.292	148

Figure 3 shows the relationship of the elastic, plastic and total strain amplitudes *versus* number of cycles to failure for Pro.I and Pro.II. The relationship curves of the elastic and plastic strain amplitudes *versus* number of cycles to failure were fitted according to Equation (1). Then parameters dependent on LCF of Pro.I and Pro.II can be obtained, as shown in Figure 3. The corresponding Manson-Coffin models are shown as follows:

$$\text{Pro.I:}\ \frac{\Delta\varepsilon_t}{2} = 0.0081\,(2N_f)^{-0.08746} + 0.05427\,(2N_f)^{-0.56959} \quad (2)$$

$$\text{Pro.II:}\ \frac{\Delta\varepsilon_t}{2} = 0.00822\,(2N_f)^{-0.0932} + 0.04663\,(2N_f)^{-0.55704} \quad (3)$$

The relationship of three-parameter power function can also be used to describe the strain-controlled LCF behavior. The relationship of total strain range to fatigue life is expressed by Equation (4) as follows:

$$N_f(\Delta\varepsilon_t - \Delta\varepsilon_0)^m = c \quad (4)$$

where $\Delta\varepsilon_0$, *m* and *c* are material constants. Fatigue limits of alloys are considered in the relationship.

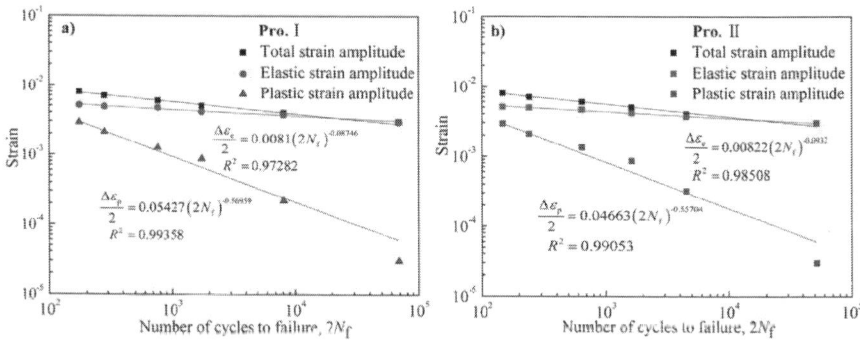

Figure 3. Total; elastic and plastic strain range *vs.* number of cycles to failure of the nickel-base alloy. (**a**) Pro.I; (**b**) Pro.II.

The relationship curves of the cycles to failure *versus* the total strain ranges for Pro.I and Pro.II are depicted in Figure 4. The relationship curves of the total strain ranges *versus* number of cycles to failure were fitted according to Equation (4). The corresponding fatigue life models are shown as follows:

$$\text{Pro.I: } N_f = 1.40603 \times 10^{-8}(\Delta\varepsilon_t - 0.00139)^{-4.43574} \quad (5)$$

$$\text{Pro.II: } N_f = 9.11575 \times 10^{-4}(\Delta\varepsilon_t - 0.00248)^{-2.27067} \quad (6)$$

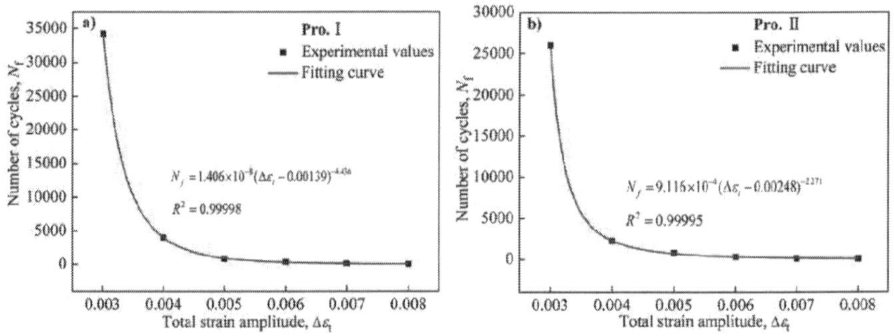

Figure 4. The reciprocal of cycles to failure *versus* the total strain ranges of the nickel-base alloy. (**a**) Pro.I; (**b**) Pro.II.

Equations (2) and (3) and Equations (5) and (6) can be used to predict the fatigue life of the alloy at 650 °C as long as the total strain amplitudes are known. The comparisons of experimental values and predicted values of 2 N_f are shown inFigure 5. The predicted effect is perfect when the fatigue life predicted points fall on the red line. It can be found that fatigue life predicted points fall basically within two fold safety factor specified dispersing band. As for Manson-Coffin relationship, the predicted effect is in good agreement with total strain amplitudes from 0.5% to 0.8% and poor for total strain amplitudes from 0.3% to 0.4%. However, as for three-parameter power function, the predicted effect is in good agreement with total strain amplitudes from 0.3% to 0.6% and poor for total strain amplitudes from 0.7% to 0.8%. The conclusion that the predicted effect of Manson-Coffin relationship is more accurate than that of three-parameter power function for high total strain amplitude section can be drawn.On the contrary, the predicted effect of three-parameter power function is more accurate than that of Manson-Coffin relationship for low total strain amplitude section. The relevance of experimental values and predicted values is used to evaluate the predicted effect of fatigue life model in the engineering. In order to evaluate fatigue life prediction method for above two kinds of fatigue life models, an error analysis method is adopted. The parameters of error analysis method are as follows [15]:

$$Err = \log(N_p/N_e) \quad (7)$$

$$\overline{E} = \frac{1}{n}\sum_{1}^{n}|Err| \tag{8}$$

where Np is predicted life; Ne is experimental life; *Err* is error and $\overline{E}$ is mean error.

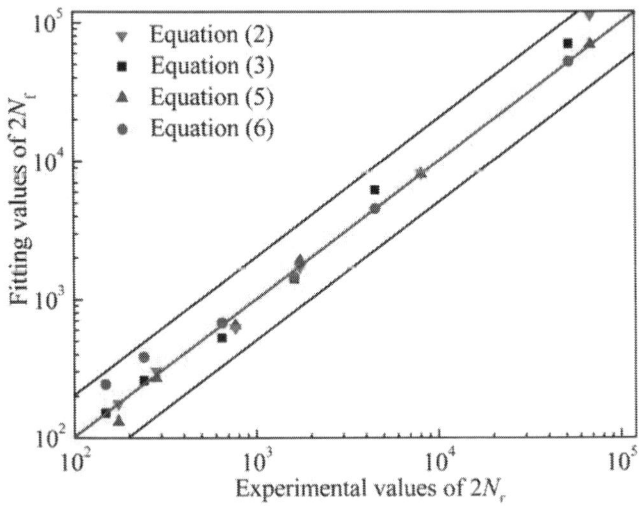

Figure 5. The comparisons of the experimental values and the fitting values.

The error analysis of the life data is presented in Table 2. Its results are consistent with the analysis of Figure 5. It can be inferred that the predicted results are consistent with experimental results. On the whole, Manson-Coffin relationship can well predict the fatigue life of the nickel-base alloy studied in this paper due to the smaller error.

Table 2. Error analysis of life prediction (%).

Heat Treatment	Manson-Coffin	Three-Parameter Power Function
Pro.I	2.83	2.79
Pro. II	2.65	6.61

3.3. Cyclic Stress Response Behavior

Cyclic stress response behavior includes cyclic hardening, softening and stability under different experimental conditions. The fundamental reason of cyclic stress response depends on the microstructure of the alloy, which includes different dislocation motions and the interactions between dislocations and γ' phases. The changes in tensile peak stress *versus*logarithm of number of cycles of Pro.I and Pro.II under total strain amplitudes of 0.4% to 0.8% are plotted in Figure 6. It can be seen that stress amplitude and cyclic stress response behavior are closely related to the strain amplitudes.

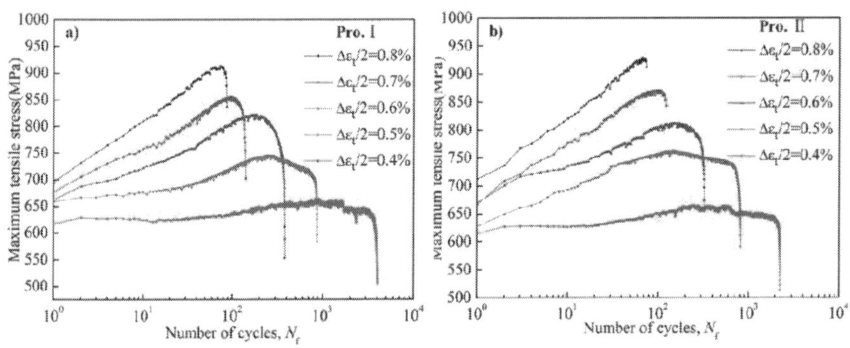

Figure 6. Cyclic stress response curve of the nickel-base alloy at 650 °C. (**a**) Pro.I; (**b**) Pro.II.

As is shown in Figure 6, the peak stress of the alloy gradually increases with the increase of total strain amplitudes in the process of LCF, from 659 MPa under low strain amplitude of 0.4% to 904 MPa under high strain amplitude of 0.8% for Pro.I and from 661 MPa under low strain amplitude of 0.4% to 910 MPa under high strain amplitude of 0.8% for Pro.II.

It can be seen in Figure 6 that cyclic hardening, softening and stability behaviors simultaneously exist in the LCF process for Pro. I and Pro. II. In terms of Pro. I, under total strain amplitudes of 0.4% and 0.5%, the alloy exhibits a continuous cyclic stability response and then cyclic hardening, followed cyclic softening till failure. Under total strain amplitudes of 0.6%, 0.7% and 0.8%, the alloy exhibits a continuous cyclic hardening response till failure and the hardening capability increases with the increase of strain amplitudes. In terms of Pro. II, the change trend of cyclic stress response curve is same with that of Pro. I except total strain amplitude of 0.5%. Under total strain amplitude of 0.5%, the alloy exhibits a continuous cyclic hardening response and then cyclic softening response till failure.

Before the final failure fracture, cyclic stress response behavior is characterized by the condition of rapid decline because of the formation of macro crack and the subsequent crack propagation till failure. It can be seen that the number of cycles for rapid decline before the final failure fracture of Pro. I are more than that of Pro. II under high strain amplitudes from 0.6% to 0.8%. It can be inferred that Pro. II causes more remarkable harming. It can also indicate that the heat treatments possess significant effect on fatigue behavior and the fatigue resistance of Pro. I is better than that of Pro. II.

3.4. Fatigue Fracture Morphology and Fracture Mechanism

Typical fatigue morphology includes fatigue source region, crack propagation region and final instant rupture region. The macroscopic and microscopic fatigue fracture morphologies of Pro.I and Pro.II are shown in Figure 7 and Figure 8, respectively, where, Zone A: fatigue source region, Zone B: crack propagation region, Zone C: final instant rupture region, Zone D: the original locations for the figures of high magnification of crack initiation site located in bottom left in Figure 7a,c,e and Figure 8a,c,e. The pictures (b), (d) and (f) in Figure 7 and Figure 8 were taken near crack propagation region.

Pro.I and Pro.II have some similar macroscopic fracture morphologies. It can be seen from the macroscopic fracture morphologies that the final instant rupture region becomes gradually larger with the increase of strain amplitudes. Moreover, there are multiple fatigue sources for the fracture morphology of the alloy after LCF at 650 °C under different strain amplitudes. It can be seen that the fatigue cracks initiate at the site of discontinuous carbides at the grain boundary from the figures of high magnification of crack initiation site. The occurrence of carbides after heat treatment results in fragile boundaries. In addition, high temperature of 650°C leads to the reduction of the strength of grain boundary, where the fatigue cracks began to initiate at the grain boundary.

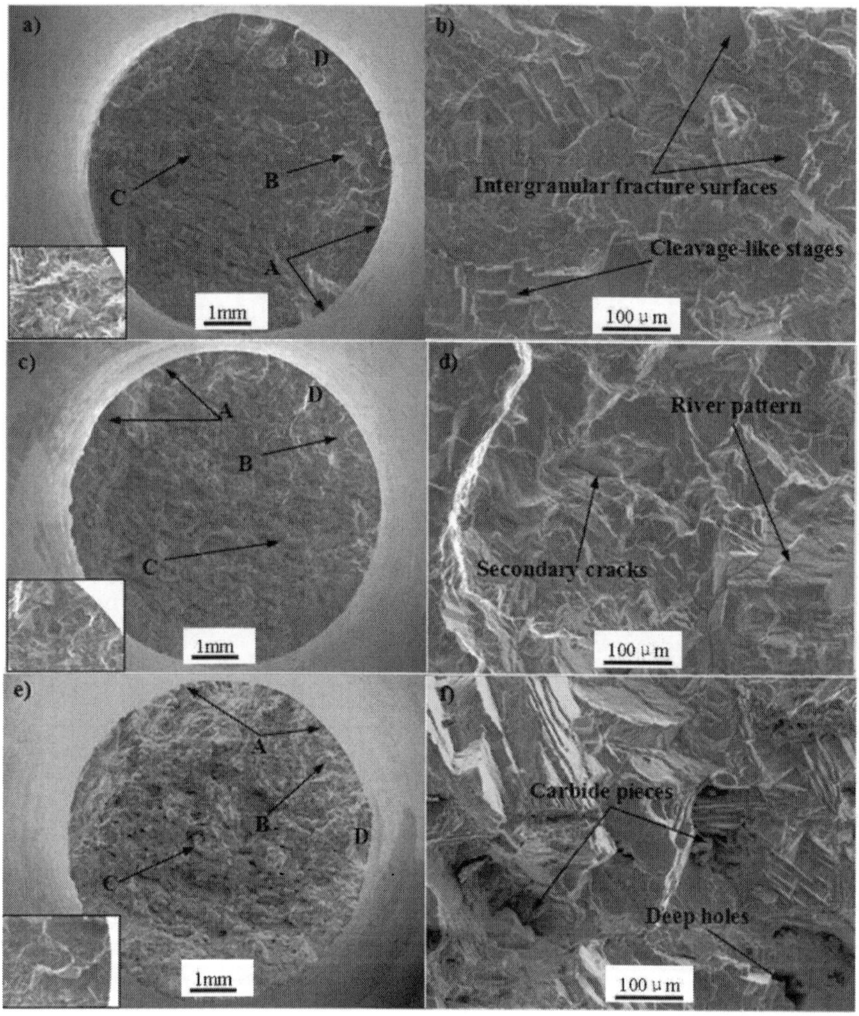

Figure 7. Fractographs of Pro.I after LCF at 650 °C. (**a**,**b**) $\Delta\varepsilon_t/2$ = 0.4%; (**c**,**d**) $\Delta\varepsilon_t/2$ = 0.6%; (**e**,**f**) $\Delta\varepsilon_t/2$ = 0.8%.

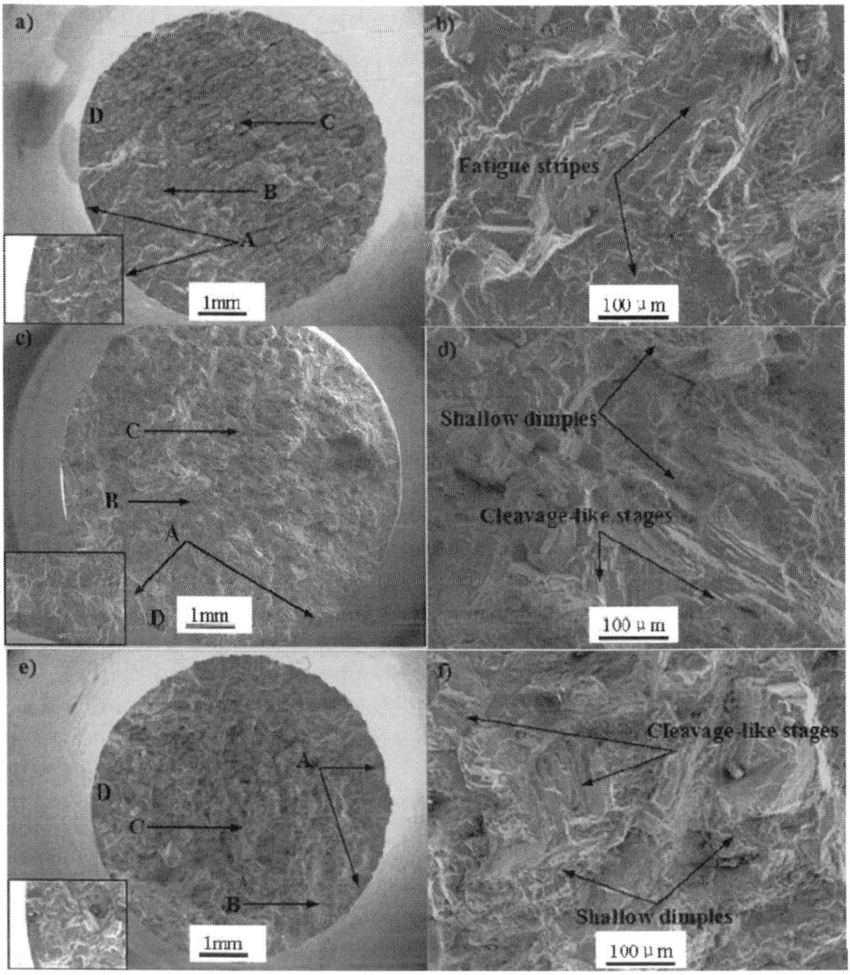

Figure 8. Fractographs of Pro.II after LCF at 650 °C. (**a,b**) $\Delta\varepsilon_t/2$ = 0.4%; (**c,d**) $\Delta\varepsilon_t/2$ = 0.6%; (**e,f**) $\Delta\varepsilon_t/2$ = 0.8%.

In terms of Pro.I, under total strain amplitude of 0.4%, intergranular fracture surfaces, secondary cracks and cleavage-like stages can be observed. The cleavage-like stage is composed of many cleavage planes. Under total strain amplitude of 0.6%, intergranular fracture surfaces, secondary cracks and river pattern can be discovered. Under total strain amplitude of 0.8%, intergranular fracture surfaces and deep holes can be found. A number of shallow dimples occur on some intergranular fracture surfaces. Carbide pieces exist in the deep holes.

In terms of Pro.II, under total strain amplitude of 0.4%, a large number of ductile fatigue stripes and a small amount of cleavage-like stages exist in fracture surface (Figure 8b). Moreover, a small amount of secondary cracks occur on the fracture surface which are parallel to fatigue stripes. Fatigue strips are caused by alternating plastic passivation opening and closing sharpening at crack tip. Under total strain amplitude of 0.6%, a small amount of the intergranular fracture and plenty of cleavage-like stages can be observed. In addition, a large number of shallow dimples can be observed. Under total strain amplitude of 0.8%, brittle dendritic fracture, a large number of cleavage-like stages and dimples can be found. Moreover, many carbide pieces exist on the fracture surface.

According to above observations, it can be concluded that the fatigue fracture mechanism of Pro.I and Pro.II during LCF process is the combined effects of brittle fracture and ductile fracture. The shallow dimples at the specimen surface indicate the occurrence of plastic deformation. The appearance of intergranular fracture surfaces and occurrence of cleavage-like stages related with the imposed strain range indicate the occurrence of brittle fracture. Hong *et al.* [16] claimed that the failure mechanism was essentially the crack initiation at the oxide-layered surface and its planar growth along <100> γ channel in LCF conditions. Yu *et al.* [17] thought that the fatigue failure mechanism was crack initiating along the slip bands that intersect the defects (micropores) along the deformed zone and shearing decohesion while the pore induces a local stress concentration. In this paper, the cracks initiate at multi-sites of free surface of the alloy (Figure 7a,c,e and Figure 8a,c,e). Then cracks propagate in some way perpendicular to the applied loading. Cracks will be subjected to obstruction when they propagate. The microscopic fatigue fracture morphologies show the interaction between dislocations and γ' phase.

The intergranular fracture surfaces occur on the fatigue fracture morphologies of Pro.I and Pro.II, which shows that the alloy produces intergranular brittle fracture due to grain boundaries weakening. The reasons of grain boundaries weakening are as follows: (1) impurity elements pile up on grain boundary; (2) the precipitates pile up on grain boundary; (3) the erosion of environmental medium; (4) high temperature. In present study, intergranular brittle fracture is caused by reasons (2) and (4). The discontinuous carbides on the grain boundary after heat treatment result in fragile boundaries in the low cycle fatigue process. At the same time, high temperature leads to the reduction of the strength of grain boundary. Thus, the fatigue properties come to decline and the intergranular fractures occur.

3.5. Dislocation Structures and Deformation Mechanism

Typical dislocation structures of Pro.I and Pro.II after LCF tests are shown in Figure 9 and Figure 10, respectively. The values of plastic deformation accumulated (pcum) during the test are also shown in Figure 9 and Figure 10.

In terms of Pro.I, under total strain amplitude of 0.4%, the distribution of dislocations is inhomogeneous. In some local regions dislocation cannot be observed. Dislocations are mainly distributed in the interface of γ' phase and the matrix phase. Dislocation pile-up is marked. Several parallel dislocation lines and different sizes of γ' phase can be found. It is obvious that dislocations cross through γ' phase (Figure 9a). Under middle strain amplitude of 0.6%, the number of dislocations significantly increases. Dislocation pile-up is more marked than some dislocation depleted regions (Figure 9b). Under total strain amplitude of 0.8%, dislocations present short shape. Most of the dislocations pile up on the boundary of γ' phase. A small portion of the dislocations cut into γ' phase (Figure 9c).

In terms of Pro.II, under total strain amplitude of 0.4%, dislocations tangle into a dislocation network. Dislocations generate interaction with γ' phases and dislocations cut into γ' phase (Figure 10a). Under middle strain amplitude of 0.6%, dislocation pile-up is more marked than some dislocation depleted regions. Most dislocations distribute in the interface of γ' phase and matrix phase. A small amount of dislocations cut into γ' phase (Figure 10b). Under total strain amplitude of 0.8%, the number of dislocations is relative the largest. The interaction of dislocations and γ' phases is most apparent. The locations of dislocation pile-up increase (Figure 10c).

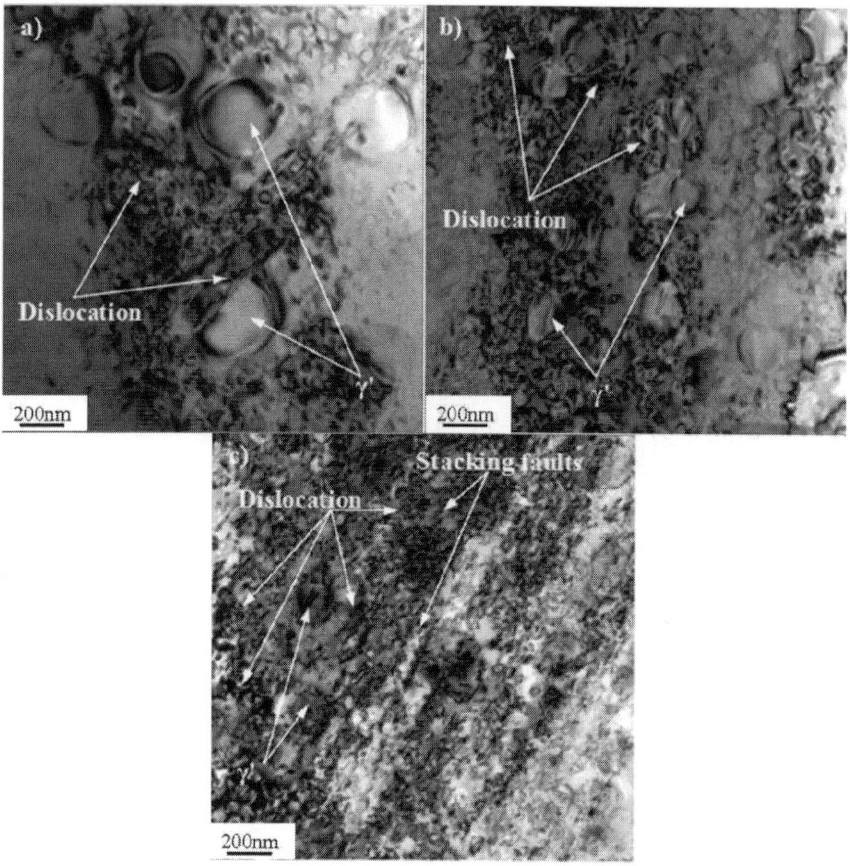

Figure 9. Typical dislocation structures of Pro.I after LCF tests at 650 °C (bright field). (**a**) $\Delta\varepsilon_t/2 = 0.4\%$, pcum = 3.562; (**b**) $\Delta\varepsilon_t/2 = 0.6\%$, pcum = 1.952; (**c**) $\Delta\varepsilon_t/2 = 0.8\%$, pcum = 1.012.

Through analysis about dislocation structures, it is obvious that the deformation of the alloy during LCF process is inhomogeneous (Figure 9a). As mentioned above, cyclic stress response behaviors include cycle hardening, softening and stability. Cyclic hardening phenomenon is caused by proliferation of dislocation hampering dislocation movement (Figure 9b). It is also associated with the interactions between dislocations and strengthening phase (Figure 10b). The cyclic softening phenomenon is due to the dislocations cutting orderly γ' phase and causing disorder, which reduces the deformation resistance of the alloy (Figure 9a and Figure 10a). It can also be caused because the growth of γ' phase reduces strengthening effect in the LCF process. Moreover, the temperature of 650 °C is high enough to provoke an aging treatment. The mobile dislocation can be trapped into a growing precipitate resulting in softening phenomenon. The cyclic stability can result from a dynamic equilibrium between the hardening effect from proliferation of dislocation hampering reciprocating motion of dislocations and the softening effect from dislocations cutting orderly γ′ phase.

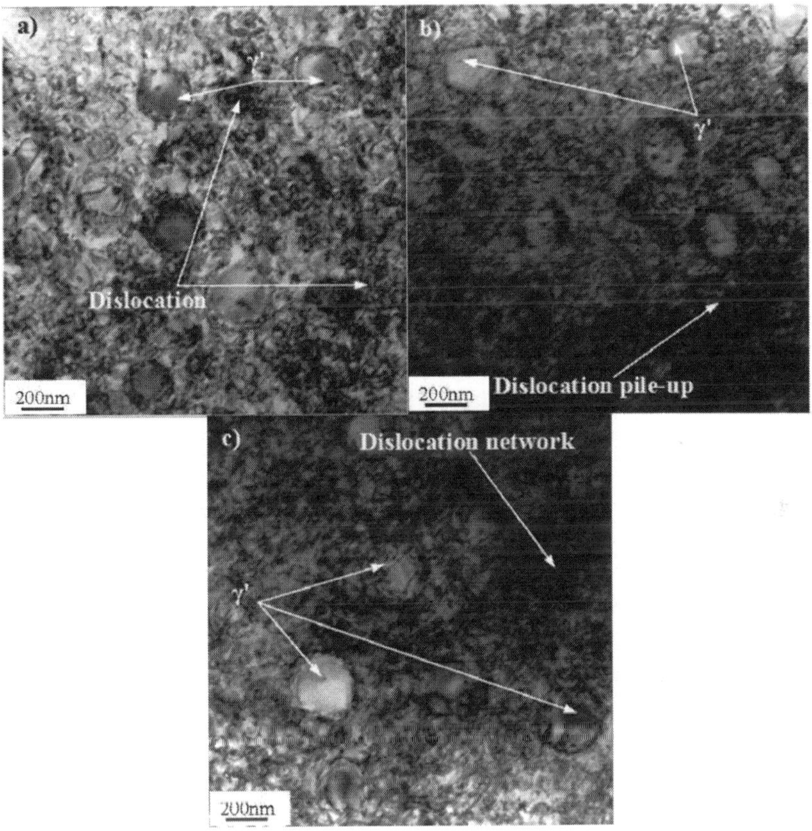

Figure 10. Typical dislocation structures of Pro.II after LCF tests at 650 °C (bright field). (**a**) $\Delta\varepsilon_t/2$ = 0.4%, pcum = 2.898; (**b**) $\Delta\varepsilon_t/2$ = 0.6%, pcum = 1.738; (**c**) $\Delta\varepsilon_t/2$ = 0.8%, pcum = 0.867.

According to above observations, it can be concluded that the deformation mechanisms of Pro.I and Pro.II during LCF process are similar. The distributions of dislocation become more homogeneous with the increase of strain amplitude. Moreover, the density of dislocation increases and dislocation pile-up becomes more marked with the increase of strain amplitude. The formation of dislocation network is caused by proliferation of dislocations (Figure 9b and Figure 10b). The proliferation of dislocations and dislocation pile-up hinder further movement of dislocations (Figure 9c and Figure 10c). The interactions between dislocations and the strengthening phase present different forms. Some dislocations run across γ' precipitates or cut into γ' precipitates (Figure 9a and Figure 10a). Some dislocations bypass γ' precipitates or bow out between γ' precipitates and form dislocation loops around γ' precipitates (Figure 10b). At lower strain levels, the precipitates are hard to deform and dislocations mainly pile up on the boundary of γ' phase. At higher strain levels, the precipitates are easy to deform and dislocations mainly cut into γ' phase. The precipitates are more prone to be cut by dislocation with increasing strain amplitude. At lower strain levels, a small number of dislocations show that the alloy produces less plastic deformation. Dislocation pile-up is hard to reach the degree of stress concentration. Thus, cracks begin to initiate after a large number of fatigue cycles. In this case, the alloy has a long fatigue life. At higher strain levels, a large number of dislocations show that the alloy produces larger plastic deformation. A large number of dislocation pile-up results in stress concentration, which leads to the occurrence of cracks initiation. Thus, the fatigue fracture of the alloy occurs in a short cycles. In this case, the alloy has a short fatigue life. It can be obtained from Figure 9 and Figure 10 that the values of plastic deformation accumulated (pcum) decrease with the increase of total strain amplitudes under the condition of the same heat treatment procedure. It is due to a large number of fatigue cycles at low total strain amplitude, though the small peak plastic amplitude, which is shown in Table 1. The proportion of the plastic strain in total strain gradually increases with increasing total strain. It increases from 1% to 36.25% for Pro.I and from 1% to 36.5% for Pro.II from the total amplitude of 0.3% to amplitude of 0.8%, which shows that plastic strain accounts for a small proportion in the total strain. It can be found that the increasing proportion of the plastic strain in total strain leads to the reduction of fatigue life for the alloy. It also indicates that the valucs of plastic deformation accumulated are associated with plastic amplitude. During the LCF process, the smaller plastic strain amplitude, the greater the values of plastic deformation accumulated, which represents the perfect fatigue resistance of the alloy. Moreover, the value of plastic deformation accumulated for Pro.I is larger than that for Pro.II at same total strain amplitude. It indicates that the nickel-base superalloy possesses more excellent fatigue resistance through the heat treatment of Pro.I than that of Pro.II.

4. CONCLUSIONS

In this paper, two kinds of heat treatment procedures were undertaken to investigate the effect of heat treatment on microstructure and LCF behaviors of a nickel-base superalloy. With the wide application of nickel-base superalloys, research on the fatigue behaviors of hot-end components such as turbine disks have become a research hotspot. In actual service process, improvement in fatigue performance is of great significance. Proper heat treatment processes significantly improv the mechanical properties of superalloys especially fatigue resistance through changing the internal organizational structure. Thus, hot-end components with perfect fatigue performance can serve safely and reliably. It is meaningful to improve the production and application of nickel-base superalloys as well as the life of hot-end components through investigating the effects of heat treatment process on LCF behaviors. The main results drawn from this study are as follows:

(1) Thc heat treatment process plays an important role in microstructure and fatigue behavior. The fatigue resistance of Pro.I and Pro.II significantly drops with the increase of strain amplitude. The value of plastic deformation accumulated for Pro.I is larger than that for Pro.II at same total strain amplitude, which shows that the fatigue resistance performance of Pro.I is better than that of Pro.II.

(2) The predicted effect of Manson-Coffin relationship is more accurate than that of three-parameter power function for high total strain amplitude section. On the contrary, the predicted effect of three-

parameter power function is more accurate than that of Manson-Coffin relationship for low total strain amplitude section.

(3) The fatigue fracture mechanism of Pro.I and Pro.II during LCF process is the combined effects of brittle fracture and ductile fracture. The main deformation mode of the alloy is inhomogeneous deformation and the different interactions between dislocations and γ' phases.

ACKNOWLEDGMENTS

This study was financially supported by the National Natural Science Foundation of China (Grant No.51575129, 51505101), Shandong Province Higher Educational Science and Technology Program (Grant No. J15LA51), and China Postdoctoral Science Foundation (Grant No. 2015M571407).

AUTHOR CONTRIBUTIONS

The paper was written under the direction and supervision of Peng Zhang and Chuanjie Wang. Peng Zhang and Qiang Zhu are responsible for the calculation, analysis and writing of this work. Gang Chen and Heyong Qin are responsible for the calculation and revision of this work.

CONFLICTS OF INTEREST

The authors declare no conflict of interest.

REFERENCES

1 Harrison, W.; Whittaker, M.; Williams, S. Recent advances in creep modelling of the nickel base superalloy, alloy 720Li. *Materials* **2013**, *6*, 1118–1137.

2 Leo Prakash, D.G.; Walsh, M.J.; Maclachlan, D.; Korsunsky, A.M. Crack growth micro-mechanisms in the IN718 alloy under the combined influence of fatigue, creep and oxidation. *Int. J. Fatigue* **2009**, *31*, 1966–1977.

3 Acharya, M.V.; Fuchs, G.E. The effect of long-term thermal exposures on the microstructure and properties of CMSX-10 single crystal Ni-base superalloys. *Mater. Sci. Eng. A* **2004**, *381*, 143–153.

4 Ott, M.; Mughrabi, H. Dependence of the high-temperature low-cycle fatigue behaviour of the monocrystalline nickel-base superalloys CMSX-4 and CMSX-6 on the γ/γ′-morphology. *Mater. Sci. Eng. A* **1999**, *272*, 24–30.

5 Brien, V.; Décamps, B. Low cycle fatigue of a nickel based superalloy at high temperature: Deformation microstructures. *Mater. Sci. Eng. A* **2001**, *316*, 18–31.

6 Shyam, A.; Milligan, W.W. Effects of deformation behavior on fatigue fracture surface morphology in a nickel-base superalloy. *Acta Mater.* **2004**, *52*, 1503–1513.

7 Lu, Y.L.; Chen, L.Y.; Wang, G.Y.; Chen, L.Y.; Wang, G.Y.; Benson, M.L.; Liaw, P.K.; Thompson, S.A.; Blust, J.W.; Browning, P.F.; *et al.* Hold time effects on low cycle fatigue behavior of HAYNES 230® superalloy at high temperatures. *Mater. Sci. Eng. A* **2005**, *409*, 282–291.

8 Marchionmi, M.; Osinkolu, G.A.; Onofrio, G. High temperature low cycle fatigue behaviour of UDIMET 720 Li superalloy. *Int. J. Fatigue* **2002**, *24*, 1261–1267.

9 Chen, L.J.; Wang, Z.G.; Yao, G.; Tian, J.F. The influence of temperature on low cycle fatigue behavior of nickel base superalloy GH4049. *Int. J. Fatigue* **1999**, *21*, 791–797.

10 Zhang, P.; Zhu, Q.; Hu, C.; Wang, C.J.; Chen, G.; Qin, H.Y. Cyclic deformation behavior of a nickel-base superalloy under fatigue loading. *Mater. Des.* **2015**, *69*, 12–21.

11 Du, J.H.; Lü, X.D.; Deng, Q. Effect of Heat Treatment on Microstructure and Mechanical Properties of GH4169 Superalloy. *Rare Metal. Mat. Eng.* **2014**, *43*, 1830–1834.

12 James, L.A.; Mills, W.J. Effect of heat-treatment and heat-to-heat variations in the fatigue-crack growth response of Alloy 718. *Eng. Fract. Mech.* **1985**, *22*, 797–817.

13 ISO 1099:2006. In *Metallic Materials-Fatigue Testing-Axial-Force-Controlled Method*; Multiple. Distributed through American National Standards Institute; The International Organization for Standardization: Geneva, Switzerland, 2007.

14 Lord, D.C.; Coffin, L.F. Low cycle fatigue hold time behavior of cast Rene 80. *Metall Trans.* **1973**, *4*, 1647–1654.

15 Chen, X.; Song, J.; Kim, K.S. Low cycle fatigue life prediction of 63Sn-37Pb solder under proportional and non-proportional loading. *Int. J. Fatigue* **2006**, *28*, 757–766.

16 Hong, H.U.; Kang, J.G.; Choi, B.G.; Kim, I.S.; Yoo, Y.S.; Jo, C.Y. A comparative study on thermomechanical and low cycle fatigue failures of a single crystal nickel-based superalloy. *Int. J. Fatigue* **2011**, *33*, 1592–1599.

17 Yu, J.J.; Sun, X.F.; Jin, T.; Zhao, N.R.; Guan, H.R.; Hu, Z.Q. High temperature creep and low cycle fatigue of a nickel-base superalloy. *Mater. Sci. Eng. A* **2010**, *527*, 2379–2389.

CHAPTER 11

Structural transformations in austenitic stainless steel induced by deuterium implantation: irradiation at 100 K

Oleksandr Morozov[1] Volodymyr Zhurba[1], Ivan Neklyudov[1], Oleksandr Mats[1], Aleksandr Rud[2], Nikolay Chernyak[1] and Viktoria Progolaieva[1]

[1]National Science Center "Kharkiv Institute of Physics and Technology", 1, Akademichna Street, 61108 Kharkiv, Ukraine

[2]Institute for Metal Physics of NASU, 36 Academician Vernadsky Boulevard, 03142 Kharkiv, Ukraine

ABSTRACT

Deuterium thermal desorption spectra were investigated on the samples of austenitic stainless steel 18Cr10NiTi preimplanted at 100 K with deuterium ions in the dose range from 3×10^{15} to 5×10^{18} D/cm^2. The kinetics of structural transformation development in the implantation steel layer was traced from deuterium thermodesorption spectra as a function of implanted deuterium concentration. At saturation of austenitic stainless steel 18Cr10NiTi with deuterium by means of ion implantation, structural-phase changes take place, depending on the dose of implanted deuterium. The maximum attainable concentration of deuterium in steel is $C = 1$ (at.D/at.met. = 1/1). The increase in the implanted dose of deuterium is accompanied by the increase in the retained deuterium content, and as soon as the deuterium concentration attains $C \approx 0.5$ the process of shear martensitic structural transformation in steel takes place. It includes the formation of bands, body-centered cubic (bcc) crystal structure, and the ferromagnetic phase. Upon reaching the deuterium concentration $C > 0.5$, the presence of these molecules causes shear martensitic structural transformations in the steel, which include the formation of characteristic bands, bcc crystal structure, and the ferromagnetic phase. At $C \geq 0.5$, two hydride phases are formed in the steel, the decay temperatures of which are 240 and 275 K. The hydride phases are formed in the bcc structure resulting from the martensitic structural transformation in steel.

KEYWORDS

Structural transformations Austenitic stainless steel Deuterium Ion implantation TDS X-ray TEM

1. BACKGROUND

The investigation into regularities of hydrogen interaction with metals and alloys over a wide range of temperatures and pressures still remains a currently central problem in material physics from both the scientific and practical standpoints [1-9]. The hydrogen accumulation in structural and functional materials is an extremely hazardous phenomenon, which leads to hydrogen degradation of materials and to possible unforeseen equipment failures [10-14]. Degradation of materials increases due to hydrogen interaction with the whole range of crystal structure imperfections of solids such as interstitial and substitutional impurities, vacancies and their complexes, dislocations and their pileups, subgrain and grain boundaries, and phase components [11,15,16].

Vehoff [17] reviewed the interaction of hydrogen with defects in metals. He states that hydrogen concentrations far in excess of the lattice concentration cannot be produced by elastic interactions alone: the amount of hydrogen in a so-called Cottrell atmosphere around a dislocation is small. In contrast, chemical interaction or trapping at preferred lattice sites such as grain boundaries, particle-matrix interfaces, or dislocation cores can lead to a rise in local hydrogen concentration that can span many orders of magnitude. The respective concentration rise is highly dependent on the local chemical and/or structural arrangement [18-23].

Temperature, mechanical, radiation, and implantation effects give rise to additional permanent structural changes in metals, alloys, and steels. These are induced phase transformations, an increase in the density of dislocations and vacancies, pair production. All these factors enhance the hydrogen accumulation and lead to the loss of plasticity, and later on, to the failure [11,12,24-29].

Hydrogenation of the face-centered cubic (fcc) iron-based alloys, which constitute an array of austenitic stainless steels, can cause phase transformations: fcc (γ) $\rightarrow$ bcc (α^*) and fcc (γ) $\rightarrow$ hcp (ε) [12,30-35]. The hydrogen-induced phases are sometimes considered as pseudohydrides. It is generally assumed that the role of hydrogen consists in creation of a particular stress state that triggers the phase transformation. In contrast, Vakhney and co-workers [36] showed that the phase transformation fcc (γ) $\rightarrow$ hcp (ε) was also due to a change in the electronic structure of the alloy after hydrogenation. At the same time, several studies have shown that martensite formation is responsible for hydrogen-enhanced crack growth, which results in quasi-cleavage fracture of specimens tested in hydrogen [37,38]. Gey and co-workers [39] found that the amount of strain-induced ε- and α^*- martensite in AISI 304 steel (Pittsburgh, PA, USA) was strongly dependent on the local mutual orientations of neighboring grains, i.e., texture of the γ steel.

Stainless steel is one of the most useful classes of engineering materials. For example, austenitic steel is used for manufacturing vessel internals of fission-type reactors. A wide use of austenitic stainless steels as structural fission reactor elements calls for a detailed knowledge of their behavior under conditions of radiation influence, accumulation of gas impurities (hydrogen isotopes, above all) [12-14,19,26-29,40,41].

Note that the hydrogen effect on the martensitic phase transformations in austenitic stainless steels was observed only as a result of hydrogenation by the electrochemical method or from the gaseous phase. Furthermore, the hydrogen saturation of steel was performed both during and after different structure changing effects such as torsion, plastic deformation, dilatometry, and extrusion.

As to our knowledge, there are no literature communications about the observation of martensitic phase transformations in austenitic stainless steels resulting from implantation of hydrogen isotope ions, and about ultimate concentrations of hydrogen isotope retention in steels. Apparently, this is due to steel irradiation conditions, which impede the accumulation of high concentrations of hydrogen isotopes in austenitic stainless steels (irradiation temperature, ambient pressure of hydrogen, etc.).

One of the most informative methods of investigating hydrogen behavior in metals is the thermal desorption spectrometry (TDS). The TDS technique enables one to determine temperature ranges of implanted hydrogen retention and release, to find thermoactivation parameters, and to establish quantitative characteristics of hydrogen emission-re-emission. It also shows fair correlation between the thermoactivated hydrogen release spectra and the metal-hydrogen phase diagrams [42-44].

The present paper reports the results from studies on implanted deuterium-induced structural transformations in austenitic stainless steel 18Cr10NiTi as functions of implanted deuterium doses at a temperature of 100 K. This temperature provides a low diffusion mobility of deuterium, and hence, a high deuterium concentration in the implantation layer. The structural transformation kinetics has been traced from the deuterium thermodesorption spectra versus implanted deuterium concentrations. The important methodical advantage of the present study lies in the fact that the implantation of the assigned dose of deuterium and the measurements of temperature ranges of its desorption were performed in one and the same setup (*in situ*).

2. METHODS

The TDS technique has been used to investigate the kinetics of spectrum development for deuterium desorption from austenitic stainless steel 18Cr10NiTi versus the implanted deuterium dose. In our experiments, to reduce the impact of background hydrogen being present in the samples and in the target chamber, we have used the hydrogen isotope deuterium.

The sample irradiation and the thermodesorption spectral measurements were carried out using the experimental facility 'SKIF' described in detail in ref. [45].

The samples were pre-implanted with 12 keV deuterium ions (D+2D2+, 24 keV) of current density 5 μA/cm^2 in the dose range from 3×10^{15} to 5×10^{18} D/cm^2 at the sample temperature $T_{\text{irr.}} =$ 100 K. The low temperature was chosen to restrict the diffusion mobility of deuterium in the samples and to investigate its behavior in a wide range of concentrations produced in the implantation layer.

The samples were homogenized at 1,350 K for 30 min in a vacuum of 5×10^{-5} Pa. The austenitic stainless steel 18Cr10NiTi samples (Table 1), measuring $10 \times 5 \times 0.3$ mm^3, were fastened to the same-steel heaters, measuring $40 \times 5 \times 0.3$ mm^3. The radiation area was $S = 0.3$ cm^2. Prior to irradiation, the samples underwent short-time annealing (1 min.) at 1,300 K in order to degas them and to remove impurities from their surface. After irradiation, the samples implanted to the pre-assigned dose were heated in the same measurement chamber at a rate of 3.5 K/s to a temperature of 1,700 K, with simultaneous registration of the D+2D2+ ion desorption spectrum (4 amu). The heating of samples started immediately after the ion beam was switched off. The temperature was measured by the tungsten-rhenium thermocouple WRe 5/20 fixed to the sample. The total amount of deuterium released from the sample was determined from the area lying under the gas release curve.

Table 1 Chemical composition of austenitic stainless steel 18Cr10NiTi (weight percent)

	C	Cr	Ni	Ti	Mn	Si	P	S	Fe
18Cr10NiTi	Approximately 0.1	17.1 to 18.5	9.6 to 10.2	0.45 to 0.56	1.0 to 1.2	0.4 to 0.43	0.03	0.01	Rest

The sample structure was examined by means of transmission electron microscopy at a voltage of 100 kV. The steel samples were thinned down to a thickness of 50 to 60 μm through chemical etch polishing, following which deuterium ions were implanted to the given dose. Thereafter, disks, 3 mm in diameter, were cut out from the obtained foil (the diameter was specified by the sample holder for the microscope), and a final thinning was done in the same solution.

The X-ray diffraction analysis of the samples was performed with the diffractometer HZG-4 (Seifert, Berlin, Germany) in the CuKα-radiation (β-filter).

The magnetic characteristics of 18Cr10NiTi steel and iron were measured with a bar-and-yoke permeameter. In this case, the sample in the form of a plate was placed in the axial (perpendicular to its plane) field.

3. RESULTS AND DISCUSSION

Kinetics of deuterium thermal desorption spectrum development in relation to the implanted deuterium dose

The most characteristic deuterium thermodesorption spectra for different implanted deuterium doses are shown in Figure 1. It can be seen that at low implantation doses (see Figure 1), the thermodesorption spectrum of ion-implanted deuterium represents a single peak with the maximum at $T_m = 405$ K, which we call the peak *a*. The presence of a single peak in the deuterium thermodesorption spectrum at low D concentrations suggests the conclusion that it characterizes the formation of the phase state of deuterium solid solution in the steel (see Figure 2). The appearance of the second peak in the thermodesorption spectrum points to the completion of the phase state formation of deuterium solid solution in steel. The deuterium concentration present in the implantation layer corresponds to ~2 at.% D for a dose of 1.8×10^{16} D/cm^2.

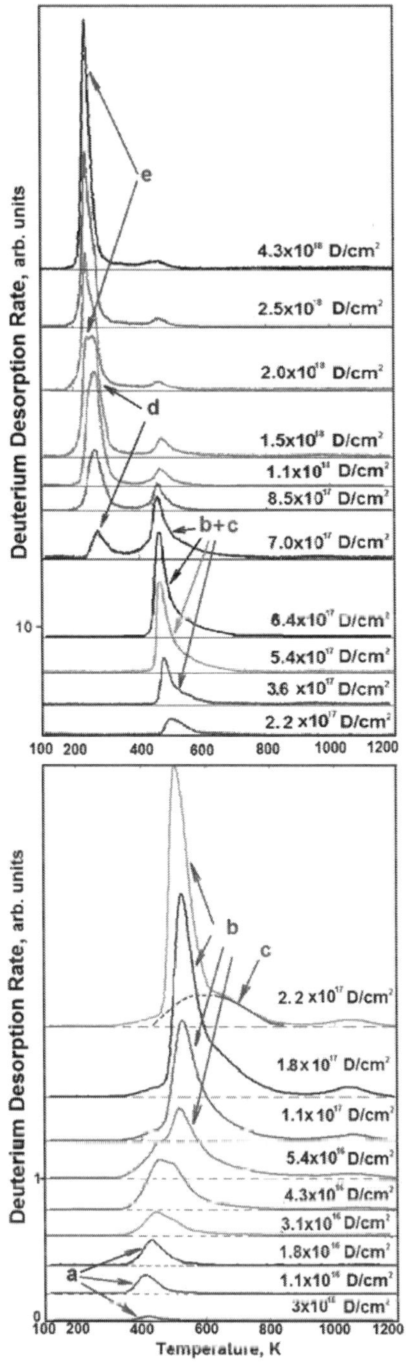

Figure 1 Deuterium thermodesorption spectra from 18Cr10NiTi steel samples implanted with different doses of deuterium ions.

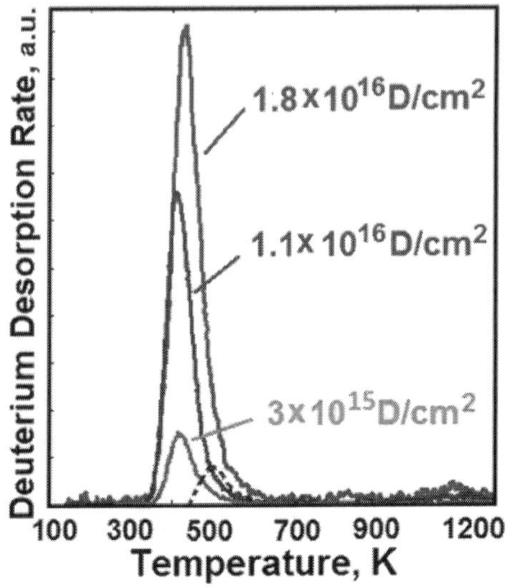

Figure 2 Deuterium thermodesorption spectra from 18Cr10NiTi steel samples implanted to doses: 3×10^{15} D/cm^2; 1.1×10^{16} D/cm^2; 1.8×10^{16} D/cm^2.

At low deuterium concentrations, the second, higher-temperature peak (peak *b*, maximum temperature $T_m = 500$ K) is practically coincident with the low-temperature peak. An increase in the implanted deuterium dose is accompanied by enhancement in the peak intensity, and by the appearance of a moderate-intensity deuterium desorption region (extended in the temperature scale) at temperatures between 500 and 800 K (peak *c*). The presence of this region indicates that apart from the fractions with discrete values of thermodesorption activation energy, there is a gas flow, the activation energy of which smoothly varies in wide temperature ranges. We identify this temperature scale-extended region of deuterium desorption as an amorphous phase state of the solid, the presence of which is necessary for structural transformations.

The excess of the implanted deuterium dose 1.8×10^{17} D/cm^2 is accompanied by a gradual decrease in the *b* peak maximum temperature, which becomes equal to 430 K at a dose of 6.4×10^{17} D/cm^2. At the same time, as the implantation dose increases, an essential change takes place in the geometric shape of this thermodesorption spectral peak. It takes on a characteristic form with a sharp, nearly vertical, front, and a slowly falling-off high-temperature part.

The observed decrease in the temperature of the peak *b* maximum can be mainly due to both the stress buildup owing to a growing concentration of implanted deuterium, and the presence of the amorphous component in the steel structure (peak *c*). This temperature region appears practically simultaneously with the second peak in the deuterium thermodesorption spectrum; it has the maximum intensity (amount of desorbed deuterium in the peak *c*) at the dose 6.4×10^{17} D/cm^2 and shows a decrease in the intensity with a further increase in the implanted deuterium dose.

At doses higher than 6.5×10^{17} D/cm^2, the deuterium thermodesorption spectrum qualitatively changes, this being manifested in the appearance of a low-temperature region of deuterium desorption, which testifies to the emergence of a new phase state; that may be considered as a hydride formation. The conclusion about hydride formation was made on the basis of our data obtained when studying the deuterium thermodesorption spectra from Pd [42] and Ti [43]. It was indicated there that in the deuterium thermodesorption spectrum, the hydride formation showed itself as the occurrence of lower-temperature peaks.

From the given TDS spectra, it can be seen that at the irradiation dose 7×10^{17} D/cm^2, apart from the previously observed peaks *b* and *c*, there is only one, strongly blurred peak centered at 275 K, which is due to the hydride formation (peak *d*). With an increasing dose, this peak grows in its amplitude, and the spectrum exhibits one more peak with the maximum at 240 K (peak *e*). At 1.5×10^{18} D/cm^2, the amplitudes of these low-temperature peaks become practically equal (see Figure 1). A further increase in the implanted deuterium dose leads to a rapid increase in the intensity of the lowest-temperature peak of the deuterium thermodesorption spectrum, with the result that the peak becomes prevailing. In this case, the intensities of the peaks *b* and *d* decrease. The presence of clearly defined peaks *d* and *e* in the deuterium thermodesorption spectra suggests an important conclusion that in the steel-deuterium system, there are two hydride phases degrading at temperatures of 240 and 275 K. It is also obvious from Figure 1 that the lowest-temperature hydride phase (peak *e*) is formed due to both the newly implanted deuterium and the earlier formed higher-temperature phases (peaks *d*, *b*, and *c*).

Note that the formation of low-temperature regions of deuterium desorption is accompanied by the appearance of the deuterium desorption region, extended in the temperature scale in the range between 270 and 430 K. This provides additional supporting evidence that the amorphous phase state of the solid is necessary for the realization of structural changes.

Amount of deuterium in different phase states of the deuterium-steel system

Using the deuterium thermodesorption spectra from the 18Cr10NiTi steel samples exposed to different doses, we have plotted the total amount of deuterium *C*(*F*) desorbed from the sample as a function of the radiation dose *F*. This function is presented in Figure 3. The figure also shows the dose dependences of the amount of deuterium desorbed from three most intense peaks observed in the TDS spectra.

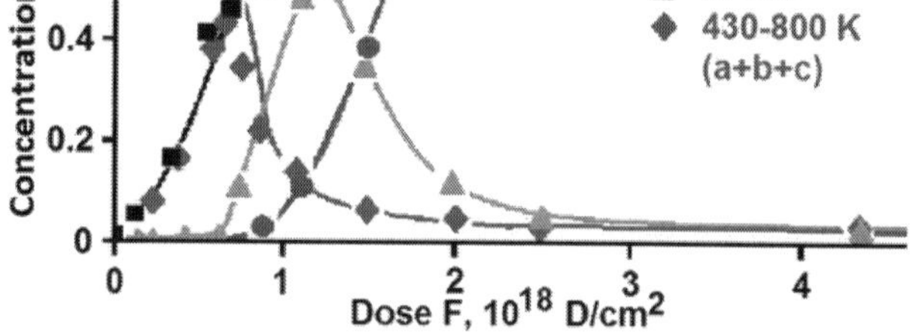

Figure 3 Deuterium content in each individual peak of the spectrum and the integral quantity of desorbed deuterium as functions of radiation dose for austenitic steel.

The linear dependence of the amount of implanted deuterium on the irradiation dose value persists until the dose $F = 1.3 \times 10^{18}$ D/cm^2 is reached. Then we observe an abrupt departure from the linearity and the tendency for attaining saturation at a dose of 1.5×10^{18} D/cm^2, though the true saturation is not attained. In this case, the excessive deuterium is mostly desorbed from steel, and only a small part of deuterium reaches the steel deuteride-steel interface, thereby providing the deuteride penetration deep into the sample, and also, the increase in the amount of trapped deuterium.

The concentrations of implanted deuterium were estimated with due regard for both the amount of metal atoms in the implanted layer and the amount of implanted deuterium. In this case, it was assumed that implanted deuterium had low diffusion mobility in the steel cooled down to 100 K, and practically all deuterium was in the implantation layer. The estimation has shown that saturation of 18Cr10NiTi steel with deuterium is attained at deuterium concentration C (at.D/at.met.) ≈ 1.0. When plotting the $C(F)$ curve, it was taken into account that the steel surface reflection coefficient for the D^+ ions of energy 12 keV per deuteron was about 10% [46].

At the initial stage of implantation dose increase, we observe the formation of the phase state of deuterium solid solution in steel (peak *a* with the maximum temperature 405 K at $C = 0.02$ at.D/at.met. = 2 at.% deuterium), and also, the emergence of two regions of deuterium desorption in the temperature range between 430 and 800 K (see Figure 1, peaks *b* and *c*), the total concentration of which reaches $C = 0.5$ at.D/at.met.

The excess of deuterium concentration in steel $C \geq 0.5$ at.D/at.met. gives rise to the low-temperature region of deuterium desorption. First, the hydride phase is formed, which degrades at 275 K (peak *d* in Figure 1), and then, with some delay in the dose, the second hydride phase is formed, with the degradation temperature of 240 K (peak *e*). The increase in the amount of deuterium in the hydride phase that degrades at 275 K (peak *d* in Figure 1) continues until the concentration $C = 0.5$ at.D/at.met. is attained. At that, the intensity of the earlier formed higher-temperature phases (peaks *b* and *c*) decreases.

The formation of the lowest-temperature hydride phase that decays at 240 K (peak *e*) occurs due to both the newly implanted deuterium and the earlier produced higher-temperature phases (peaks *d*, *b*, and *c*). The highest deuterium concentration in the low-temperature phase of steel is attained at $C = 0.9$, this approaching the stoichiometry at.D/at.met. = 1/1.

On the nature of deuterium retention in the 18Cr10NiTi steel

From the analysis of the development kinetics of the deuterium thermodesorption spectrum versus the dose of implanted deuterium, it follows that after formation of the phase state of deuterium solid solution in the 18Cr10NiTi steel, a higher-temperature phase state is formed, which is not typical of hydride-forming metals. Note that the hydride-forming metals are characterized by the formation of hydrides having a lower decay temperature of the phase state of deuterium solid solution in the metal [42,43]. This manifests itself in the thermodesorption spectra of deuterium by the emergence of lower-temperature peaks. This result points to different nature of deuterium retention in the austenitic stainless steel 18Cr10NiTi and in the hydride-forming metals. The data suggest an important conclusion that the austenitic steel, having the fcc structure, forms no hydrides.

It is then quite logical to assume that the phase state with the maximum temperature 500 K (peak *b*, Figure 1) represents the kinetics of austenitic stainless steel lattice occupation without formation of a chemical compound (hydride). From the data on the amount of deuterium desorbed (see Figure 2) from the steel in this phase state, one can obtain the deuterium concentration value, which corresponds to $C = 0.5$ at.D/at.met. = 1/2.

Based on these data, the following scheme of deuterium atoms arrangement in the fcc lattice of steel can be offered: two deuterium atoms are arranged diagonally in two out of eight tetrahedral sites, producing a strong distortion along the (111) axis of the fcc lattice (see Figure 4). With occupation of practically all elementary cells of the implantation steel layer with deuterium, and with attainment of the concentration $C \geq 0.5$ at.D/at.met. = 1/2, a shear martensitic structural transformation and the relaxation of implanted deuterium-induced stressed state take place.

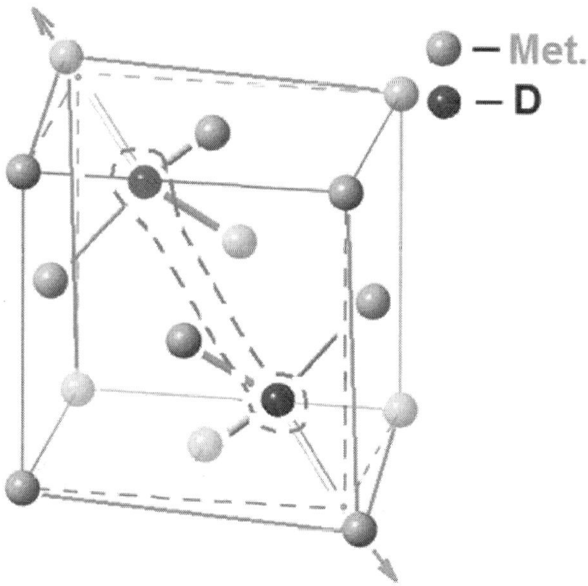

Figure 4 Crystalline structure of austenitic stainless steel with deuterium at $\boldsymbol{C} = 0.5$ at.D/at.met.

The electron microscope studies of the samples (see Figure 5) have confirmed the realization of the process of shear martensitic structural transformation on attainment of the concentration $C \geq 0.5$ at.D/at.met.

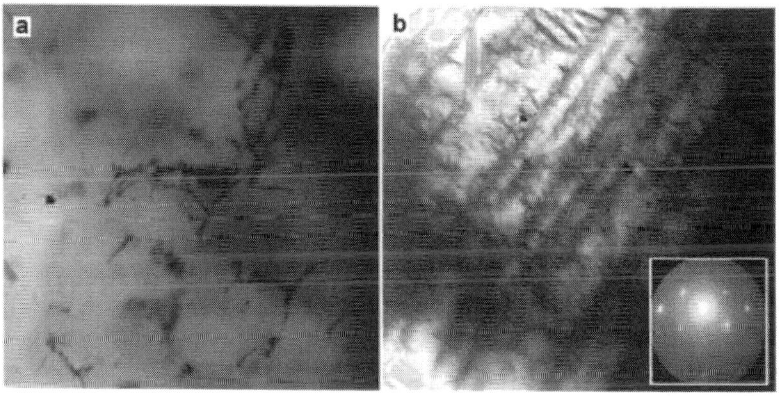

Figure 5 Structure of 18Cr10NiTi steel: (a) initial and (b) after deuterium ion irradiation to a dose of 8.2×10^{17} D/cm^2. ×90,000 magnification. The inset in **(b)** shows the electron diffraction pattern.

The electron-microscopical images of the initial samples exhibit the peculiar to the austenitic stainless steel large-size crystallites, and the randomly distributed inclusions scattered all over the sample (Figure 5a). As it is obvious from Figure 5b, in the surface layer of the steel irradiated with deuterium ions, there are austenites, the needles of α-martensite and ε -interlayer, the presence of which testifies to the actual shear polymorphous transformations realized through the intermediate ε-hcp phase: $\gamma \rightarrow \varepsilon \rightarrow \alpha$.

X-ray diffraction studies of similar samples (see Figure 6) also provide evidence for the realization of the process of the shear martensitic structural transformation on attainment of the deuterium concentration $C \geq 0.5$ at.D/at.met., namely, the appearance of the body-centered cubic (bcc) structure has been found against the background of intense lines of the fcc structure of the 18Cr10NiTi steel. A low bcc phase intensity is due to the fact that the approximately 150 nm thick implantation layer, where phase changes have taken place, is substantially smaller in thickness than the sample under study.

An additional verification of the realization of the process of shear martensitic structural transformation on attainment of the deuterium concentration $C \geq 0.5$ at.D/at.met. is provided by measurements of the magnetic characteristics of steel. For comparative tests, samples from the 18Cr10NiTi steel and Armco iron (AK Steel International, Stevenage, Great Britain) were prepared as plates nearly of the same size, the difference being no more than 1%. The measurements were carried out using the bar-and-yoke permeameter. In this case, the sample was arranged in the axial (perpendicular to its plane) field. Figure 7 depicts the magnetization curves ($M(H)$) of the samples under study. The initial 18Cr10NiTi steel samples, which underwent austenitizing annealing, show up no hysteresis dependence typical of ferromagnetic materials. The examination of steel samples implanted with deuterium ions to a dose of 1×10^{18} D/cm^2 at a temperature 100 K has revealed the appearance of the hysteresis dependence typical of a ferromagnetic (see Figure 7, curve a). To estimate the ferromagnetic phase concentration resulting from deuterium implantation, the experiment was made to investigate the Armco iron magnetization (see Figure 7, curve b). The comparative analysis has led to the conclusion that the ferromagnetic phase formed in the 18Cr10NiTi steel during deuterium implantation is no more than 2% to 3% of the sample volume (with due regard for thc experimental error). The low intensity of the registered ferromagnetic phase is due to the fact that the implantation layer, where structural-phase transformations of the fcc structure (with no ferromagnetic phase) into the bcc structure (having hysteresis dependence typical of ferromagnetics) took place, has the thickness considerably smaller than that of the sample under study.

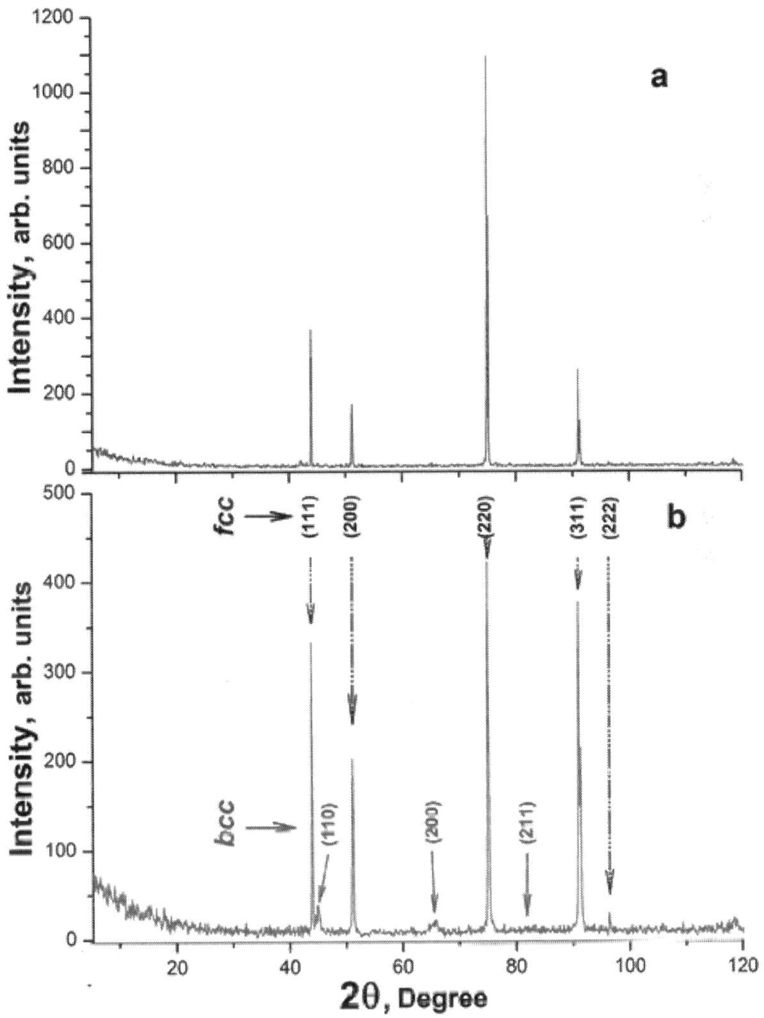

Figure 6 Diffractograms of 18Cr10NiTi steel: (a) initial and (b) irradiated with deuterium ions to 8.2×10^{17} D/cm^2.

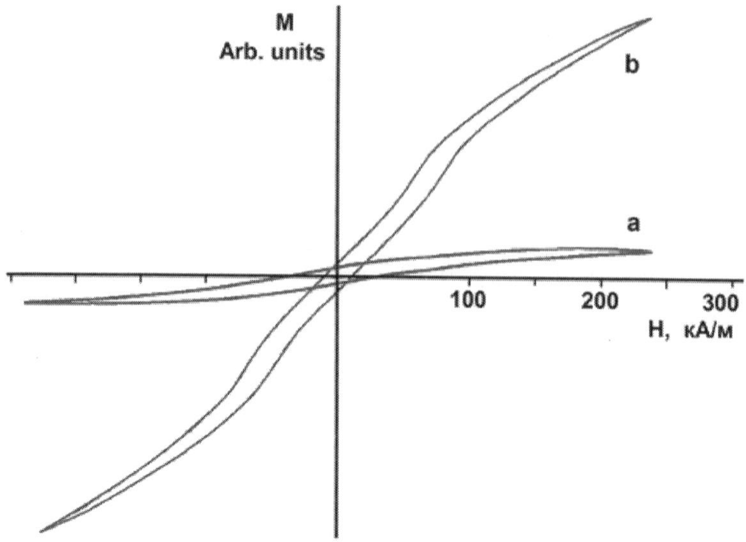

Figure 7 Magnetization curves of the samples, $M(H)$: (a) 18Cr10NiTi steel implanted with deuterium ions to 1×10^{18} D/cm^2 at 100 K;(b) Armco iron sample.

Nonetheless, the estimates have shown that the ferromagnetic phase formed in the 18Cr10NiTi steel during deuterium implantation exceeds the implantation layer thickness by nearly an order of magnitude. Hence, the process of the shear martensitic structural transformation on attainment of the deuterium concentration $C\geq0.5$ at.D/at.met. is realized to the depth exceeding the implantation layer thickness.

Notice that the bcc structure formation in the 18Cr10NiTi steel occurs simultaneously with the appearance of low-temperature peaks. These data permit us to conclude that the hydrides in the austenitic stainless steel are produced only after the bcc structure inclusions are formed in the steel, and, as a consequence, it follows that the structure of the steel hydride sublattice represents the bcc structure.

Based on the belief that deuterium atoms in the bcc crystal lattice of steel are in the charged state, are in octahedral sites, and at regular intervals, one can elaborate the arrangement of deuterium atoms in the hydride phase of steel with the deuterium concentration $C=1/2$ at.D/at.met. = 0.5 (see Figure 8a, peak d in Figure 1). A further increase in the implanted deuterium concentration is accompanied by the formation of the lower-temperature hydride phase in the same crystal lattice, where the hydride structure with deuterium concentration $C = 0.5$ has already formed. This phase is formed due to both the newly implanted deuterium, and the earlier formed higher-temperature hydride phase (peak *d*). In this case, the local concentration of deuterium in the steel-lattice crystal structure reaches $C = 1$ at.D/at.met. (see Figure 8b, peak*e* in Figure 1). The decay of this lowest-temperature hydride phase in the deuterium thermodesorption spectrum is registered as a single peak. Note that in the Pd-D system, two stable hydride sublattices in palladium are also formed [42].

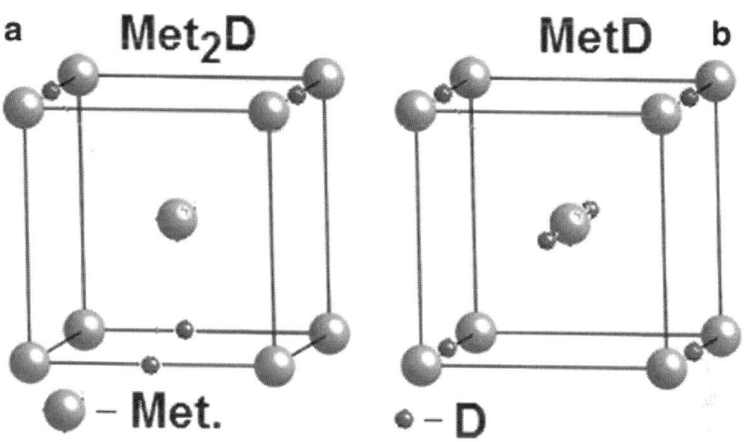

Figure 8 Crystalline structure of austenitic stainless steel with deuterium concentrations (a) $C = 0.5$ at.D/at.met. and (b) $C = 1$ at.D/at.met.

The important conclusion that follows from the present experimental data is that *the appearance of a low-temperature region of deuterium desorption from steel is connected with the α-phase formation*, i.e., the formation of the bcc structure typical of the low-temperature phase of iron, being the principal chemical component of the austenitic stainless steel.

The second important conclusion is that the hydride phase formation occurs not until the bcc structure has developed. Then, it follows that, firstly, in the fcc structure of the 18Cr10NiTi steel hydrides are not formed, and secondly, the metal sublattice of the hydride phases of 18Cr10NiTi steel has the bcc structure.

Notice that the formation of bcc structure inclusions in the austenitic stainless steel has been observed using the TEM technique, the X-ray diffraction method; the measurements of magnetic properties (see Figures 4, 5, and 6). These data provide an additional support for the fact that the peaks in the deuterium thermodesorption spectra are related with the phase transformations in metals.

The results of the studies have demonstrated that the amount of the retained deuterium is dependent on the structural-phase state of the steel.

At saturation of austenitic stainless steel 18Cr10NiTi with deuterium through ion implantation at a temperature of 100 K, there occur structural-phase changes, dependent on the implanted deuterium dose. At low implantation doses, the phase state of deuterium solid solution in the steel is formed, the deuterium concentration being 2 to 4 at.%. With an increase in the implanted deuterium dose, the concentration of the retained deuterium increases. The ultimate concentration of deuterium retention in the fcc crystal lattice of austenitic steel attains $C = 0.5$ at.D/met. Deuterium occupies interstitial fcc lattice sites of the austenitic steel.

A further increase in the implantation dose leads to the martensitic structural transformation in the steel with the result that the fcc crystal lattice of the steel changes into bcc lattice (this has been registered by the methods of TEM, X-ray diffraction, and by measurements of magnetic properties). The deuterium concentration increase in the bcc lattice of steel continues with an increase in the implantation dose up to$C = 1$ at.D/met. In this case, deuterium is found in the octahedral interstitial bcc lattice sites of the steel, occupying half of the available ones.

The TDS technique was used to determine the deuterium concentration in the phase states, and also to trace the process of structural transformations in the steel (see Figures 2 and 3).

4. CONCLUSIONS

At saturation of the austenitic stainless steel with deuterium through ion implantation at a temperature of 100 K, structural-phase changes take place, depending on the implanted deuterium dose.

The highest attainable concentration of deuterium in steel has been determined to be $C = 1$ at.D/at.met.

The increase in the implanted dose of deuterium is accompanied by the increase in the retained deuterium content, and as soon as the deuterium concentration attains $C \approx 0.5$, the process of shear martensitic structural transformation in steel takes place. It includes the formation of bands, bcc crystal structure, and the ferromagnetic phase.

At $C \geq 0.5$, two hydride phases are formed in the steel, the decay temperatures of which are 240 and 275 K. The hydride phases develop in the steel bcc structure, which results from the martensitic structural transformation.

The present data have confirmed that the peaks observed in the deuterium thermodesorption spectra are related to the phase transformations in metals. The kinetics of structural transformations development in the implantation layer of steel has been traced from the spectra of deuterium thermal desorption as a function of implanted deuterium concentration.

ACKNOWLEDGEMENTS

This study has been done with financial support to the program at the National Academy of Sciences of Ukraine 'Fundamental investigation materials science problems of functional materials' (subject no. III-4-11, ID 0111U009715, National Science Center 'Kharkiv Institute of Physics and Technology').

COMPETING INTERESTS

The authors declare that they have no competing interests.

AUTHORS' CONTRIBUTIONS

OM and VI conceived and designed the experiments, and analyzed and summarized the obtained results. IM supervised the work and reviewed the manuscript. OV carried out the TEM studies. AD carried out the X-ray measurement. NA carried out the magnetic characteristic studies. VO participated in the TDS studies and analysis TDS spectra. All authors read and approved the final manuscript.

REFERENCES

1 Alefeld G, Völkl J. Hydrogen in metals I. Verlag, Berlin, Heidelberg, New York: Springer; 1978. p. 426.

2 Alefeld G, Völkl J. Hydrogen in metals II. Verlag, Berlin, Heidelberg, New York: Springer; 1978. p. 387.

3 Wipf H. Hydroqen in metals III. Properties and applications. Berlin, Heidelberg, New York: Springer; 1997. p. 348.

4 Gel'd PV, Ryabov RA, Kodes ES. Hydrogen and metal structure imperfections (Russian). Moskow: Metallurgiya; 1979. p. 221.

5 Ageyev VN, Bekman IN, Gol'tsov VA. Hydrogen interaction with metals (in Russian). Moskow: Nauka; 1987. p. 296.

6 Lewis FA, Aladjem A. Hydrogen in metal systems. II. Pfaffikon, Switzerland: Trans Tech Publications; 2000. p. 518.

7 Varin RA, Czujko T, Wronski ZS. Nanomaterials for solid state hydrogen storage. New York: Springer; 2009. p. 338.

8 Zuttel A, Borgschulte A, Schlapbach L. Hydrogen as a future energy carrier. Weinheim: WILEY-VCH Verlag GmbH & Co. KGaA; 2008. p. 483.

9 Broom DP. Hydrogen storage materials. The characterization of their storage properties. London: Springer; 2011. p. 258.

10 Kolachev BA. Hydrogen embrittlement of metals (in Russian). Moskow: Metallurgiya; 1985. p. 216.

11 Gol'tsov VA. Progress in hydrogen treatment of materials. Donetsk–Coral Gables: Kassiopeya Ltd; 2001. p. 544.

12 Borchers C, Michler T, Pundt A. Effect of hydrogen on the mechanical properties of stainless steels. Adv Eng Mater. 2008;10:11–23.

13 Touge M, Miki T, Ikeya M. Effects of X-ray irradiation on hydrogen-induced phase transformations in stainless steel. Metall Trans. 1983;14A:151–2.

14 Neklyudov IM, Ozhigov LS, Shilyayev BA, Laptev IN, Parkhomenko AA, Morozov AN, et al. Hydrogen in stainless steel in-vessel components of the WWER-1000 reactor (in Russian). Voprosy At Nauki I Tekhniki Series: Fizika radiats povr i radiats materialoved. 2003;83:47–50.

15 Myers SM, Baskes MI, Birnbaum HK, Corbett JW, DeLeo GG, Estreicher SK, et al. Hydrogen interactions with defects in crystalline solids. Rev Modern Phys. 1992;64:559–617.

16 Gavriljuk VG. Austenite and martensite in nitrogen-, carbon- and hydrogen containing iron alloys: similarities and difference. Mater Sci Eng A. 2006;438–440:75–9.

17 Vehoff H. Hydrogen related material problems. In: Wipf H, editor. Hydrogen in metals III. Berlin, Heidelberg, New York: Springer; 1997. p. 215–78.

18 Čížek J, Procházka I, Becvár F, Kužel R, Cieslar M, Brauer G, et al. Hydrogen-induced defects in bulk niobium. Phys Rev. 2004;69B:224106.

19 Yoshida N, Ashizuka N, Fujiwara T, Kurita T, Muroga T. Radiation damage and deuterium trapping in deuterium ion irradiated austenitic stainless steel. J Nucl Mater. 1988;155–157:775–80.

20 Yoshida N, Kurita T, Fujiwara T, Muroga T. Trapping of deuterium injected in austenitic stainless steel at elevated temperatures. J Nucl Mater. 1989;162–164:1082–7.

21 Čížek J, Procházka I, Brauer G, Anwand W, Mücklich A, Kirchheim R, et al. Defect studies of hydrogen-loaded thin Nb films. Appl Surf Sci. 2006;252:3237–44.

22 Gussev MN, Busby JT, Tan L, Garner FA. Magnetic phase formation in irradiated austenitic alloys. J Nucl Mater. 2014;448:294–300.

23 Shirbanda Z, Shishesaza MR, Ashrafib A. Hydrogen degradation of steels and its related parameters, a review. Phase Transit. 2011;84:924–43.

24 Bryk VV, Neklyudov IM. Regularities of dislocation structure evolution in self-organizing materials (in Russian). Voprosy At Nauki I Tekhniki Series: Fizika radiats povr i radiats materialoved. 2001;80:9–13.

25 Rudenko AG, Shilyaev BA, Voyevodin VN, Ozhigov LS. Evolution of the radiation damage materials of the reactor WWER-1000 (in Russian). Voprosy At Nauki I Tekhniki Series: Fizika radiats povr i radiats materialoved. 2008;92:78–82.

26 Neklyudov IM, Morozov AN, Zhurba VI, Kulish VG, Galitsky AG. Hydrogen isotope retention in 18Cr10NiTi steel implanted with helium ions (in Russian). Voprosy At Nauki I Tekhniki Series: Termoyadernyi sintez. 2008;2:41–6.

27 Chernov IP, Martynenko YV, Cherdantsev YP. Mutual influence of hydrogen and helium in structural materials (in Russian). Voprosy At Nauki I Tekhniki Series: Termoyadernyi sintez. 2008;2:46–50.

28 Neklyudov I, Morozov O, Kulish V, Azhazha V, Lavrinenko S, Zhurba V. The effects of helium on temperature ranges of hydrogen

isotopes retention in Hastelloy-N alloy. J Nucl Mater. 2011;417:1158–61.

29 Neklyudov IM, Morozov AN, Kulish VG, Zhurba VI, Galytsky AG, Piatenko EV. The influence of interstitial impurities on temperature ranges of deuterium retention in austenitic stainless steel. J Nucl Mater. 2009;386–388:658–60.

30 Benson Jr RB, Dann RK, Roberts J. Hydrogen embrittlement of stainless steel. Trans Metal Soc AIME. 1968;242:2199–205.

31 Teus SM, Shyvanyuk VN, Gavriljuk VG. Hydrogen-induced $\gamma \rightarrow \varepsilon$ transformation and the role of ε-martensite in hydrogen embrittlement of austenitic steels. Mater Sci Eng. 2008;497A:290–4.

32 Kishi A, Takano N. Effect of hydrogen cathodic charging on fatigue fracture of type 310S stainless steel. J Phys Conf Ser. 2010;240:012050.

33 Mine Y, Horita Z, Murakami Y. Effect of hydrogen on martensite formation in austenitic stainless steels in high-pressure torsion. Acta Mater. 2009;57:2993–3002.

34 Rozenak P, Eliezer D. Nature of the γ and γ* phases in ustenitic stainless steels cathodically charged with hydrogen. Metal Trans. 1988;19A:723–30.

35 Rozenak P, Bergman R. X-ray phase analysis of martensitic transformations in austenitic stainless steels electrochemically charged with hydrogen. Mater Sci Eng. 2006;437A:366–78.

36 Vakhney AG, Yaresko AN, Antonov VN, Nemoshkalenko VV. The effect of hydrogen on the electronic structure and cohesive properties of iron-based alloys doped by chromium and nickel. Int J Hydrogen Energy. 2001;26:453–6.

37 Tsay LW, Liu YC, Young MC, Lin D-Y. Fatigue crack growth of AISI 304 stainless steel welds in air and hydrogen. Mater Sci Eng. 2004;374A:204–10.

38 Tsay LW, Yu SC, Huang RT. Effect of austenite instability on the hydrogen-enhanced crack growth of austenitic stainless steels. Corros Sci. 2007;49:2973–84.

39 Gey N, Petit B, Humbert M. Electron backscattered diffraction study of ε/α′ martensitic variants induced by plastic deformation in 304 stainless steel. Metal Mater Trans. 2005;36A:3291–9.

40 Wilson KL, Baskes MI. Thermal desorption of deuterium implanted stainless steel. J Nucl Mater. 1978;76&77:291–7.

41 Pontau AE, Baskes MI, Wilson KL, Haggmark LG, Bohdansky J, Scherzer BMU, et al. Deuterium retention in helium-damaged stainless steel: detrapping energy. J Nucl Mater. 1982;111–112:651–3.

42 Rybalko VF, Morozov AN, Neklyudov IM, Kulish VG. Observation of new phases in Pd-D systems. Phys Lett. 2001;287A:175–82.

43 Neklyudov IM, Morozov AN, Kulish VG. Temperature ranges of hydride phase stability of the TiD system (in Russian). Materialovedeniye. 2005;№11:45–56.

44 Escobar DP, Depover T, Duprez L, Verbeken K, Verhaege M. Combined thermal desorption spectroscopy, differential scanning calorimetry, scanning electron microscopy and X-ray diffraction study of hydrogen trapping in cold deformed TRIP steel. Acta Mater. 2012;60:2593–605.

45 Ruzhitsky VV, Gribanov YA, Rybalko VF, Khazan SM, Morozov AN, Martynov IS. The multipurpose experimental facility “SKIF” (in Russian). Voprosy At Nauki I Tekhniki Series: Fizika radiats povr i radiats materialoved. 1989;51:84–9.

46 Möller W, Besenbacher F, Bottiger J. Saturation and isotope mixing during low-temperature implantations of hydrogen into metals. Appl Phys A. 1982;27:19–29.

CHAPTER 12

Corrosion Resistance of API5L X52 Carbon Steel in Sulfide Polluted Environments

Randa Abdel-Karim*, Mohamed Ali Farag, Hafiz Abdel-Azim Ahmed, Saad El-Raghy

Department of Metallurgy, Faculty of Engineering, Cairo University, Giza, Egypt

ABSTRACT

In the present work, the corrosion behavior of base and welded API5L X-52 carbon steel coupons in synthetic ground water solution NS-4 was studied. Subsequently, the morphology and composition of the corrosion product deposits were determined using scanning electron microscopy (SEM). Based on weight loss as well as Tafel's measurements, increasing the sulfide ions content above 0.4 mM, had an inhibition effect on the corrosion behavior of carbon steel. According to EDAX analyses, increasing of sulfide concentration from 0.4 to 4 mM Na_2S, the Fe/S ratio decreased from 12.02 at 0.4 mM to 2.6 at 4 mM Na_2S.

KEYWORDS

Carbon Steel, Sulfides, Weight Loss, Anodic Dissolution, Polarization

1. INTRODUCTION

The corrosion of carbon steel in sulfide-containing solutions has attracted widespread attention for many years due to its importance in several industrial processes such as oil and gas production and transport, petroleum refining, and petrochemical/chemical processing [1] .

Some variables which affect the corrosion rate of steel in soil such as water, degree of aeration, pH, redox potential, resistivity, soluble ionic species, and microbiological activity. Concerning the environmental aspect, corrosion of steel structures in soil is a relevant problem due to soil and ground water contamination caused by failure of pipelines and other structures [2] .

The presence of some pollutants, mainly sulfide, in the oil can affect the performance of mild steel in the petrochemical industry, as this phenomenon is responsible for costly economic and human loss. Sulfide pollution of ground water can occur from the industrial waste discharge, biological and bacteriological process from seaweed, sulfide reduced bacteria, or industrial waste discharge [3] [4] .

Other applications where sulfide species have produced corrosion include water production, electric power, and marine applications. In addition, the corrosion of steel in CO_2, NaCl, Na_2CO_3, $NaHCO_3$, or Na_2S solutions constitutes the natural media often confronted in oil and gas prospecting and transportation areas [1] .

The effect of sulfide on the corrosion behavior of carbon steel in natural waters is studied by many authors [5] [6] . In the present work, the electrolyte used is a synthetic solution to simulate the soil or underground water conditions (NS-4 solution). This environment considers groundwater conditions similar to the pipelines that are exposed to underground environments while used for transmission lines. The corrosion behavior of steel in soil environment is studied using the weight loss, open circuit measurements (OCP) and Tafel's extrapolation analysis. Subsequently, the morphology and composition of the corrosion product deposits are determined using scanning electron microscopy (SEM) and energy dispersive spectroscopy (EDX).

2. EXPERIMENTAL

2.1. Materials

Experiments were conducted using API5L X-52 carbon steel pipes, provided by GASCO company-Egypt, with outer diameter 168.275 mm and the thickness 6.9 mm. The chemical composition and mechanical characteristics of this steel are given in Table 1 and Table 2. The hardness of the material was 155 HVN, and the microstructure was composed of pearlite and ferrite as shown in Figure 1. For corrosion measurements, rectangular base- as well as weld-coupons with dimensions 200 × 40 mm, were prepared and were subjected to grinding using Sic paper up to grid 1200 and polishing using alumina paste. In order to investigate the effect of plastic deformation, U-bend samples were also prepared with dimensions 200 × 40 × 6.9 mm. Three different diameters rolls of 30 mm, 40 mm, and 50 mm were used in the universal testing machine to give three different strains and hence three different stresses (Table 3).

Table 1. Chemical composition of API5LX-52 pipeline steel.

C	Si	P	S	Mn	Ni	Cr	Cu	Mo	V	Al
0.21	0.22	0.015	0.008	0.56	0.029	0.04	0.064	0.01	0.003	0.005

Table 2. Mechanical characteristics of API5LX-52 pipeline steel.

Specimen	W (mm)	Thickness (mm)	Yield Strength (Mpa)	UTS (Mpa)	Strain %
Weld	18.8	6.2	n.a	567	n.a
Base Metal	11.5	6.9	355	527	25

Figure 1. Microstructure of API5L X52 base metal (4” pipeline material).

Table 3. Roll Diameter vs. strain & stress obtained.

Roll Diameter	Strain	stress
30 mm	23%	158%SMYS
40 mm	17.25%	156%SMYS
50 mm	13.8%	152%SMYS

2.2. Electrolytes

The synthetic underground water (NS-4) electrolyte used was near-neutral pH environment (pH 8.4), whose chemical composition is given in Table 4 after Parkins et al. [7] . The sulfide stock solution was prepared by dissolving ($Na_2S.9\ H_2O$) crystals in NS-4 solution. The effect of sulfide ions concentration on the corrosion behavior of carbon steel was studied in NS-4 solution containing 0.4, 2, 4, and 10 mM Na_2S with conductivity values 1456, 2230, 2680 and 5030 μS, respectively.

Table 4. Chemical composition of NS-4 synthetic ground water solution [8] .

Chemicals	g/l
$NaHCO_3$	0.483
KCl	0.122
$CaCl_2$	0.093
$MgSO_4$	0.131

2.3. Immersion Test

The corrosion performance of carbon steel coupons in synthetic underground water (NS-4), contaminated with different sulfide contents, was studied by measuring the weight loss as well as the open circuit potential (OCP) at time intervals of 7 days for 3 month. Samples were rinsed with distilled water and carefully dried before taking weight change measurements. The corresponding corrosion rates (CR) in mils per year (mpy), based on weight loss measurements, were calculated using the following equation [8] :

$$CR = 534\frac{W}{DAt} \quad (1)$$

where W is the weight loss in gm (g), D is the density of steel g/cm^3, A is the total surface area (cm^2) and t is the total exposure time in hours.

2.4. Potentiodynamic Test

The potentiodynamic test was conducted using a standard three electrodes cell. Each specimen was used as a working electrode with an exposure area of 1 cm^2. A saturated calomel electrode was used as a reference electrode and the counter electrode was a platinum electrode. The cell was connected to PARSATAT4000 potentiostat. The scanning range was set from −200 mV to +200 mV with respect to open circuit potential using a scanning rate 0.5 mV/sec. Versa Studio Software was used for drawing polarization curves, as well as Tafel's plot.

2.5. Surface Characterization

SEM (JEOL JXA-840A) was used to demonstrate the microstructure and surface morphology of the corrosion products. Local compositions were obtained using the energy dispersive spectroscopy (EDS) unit attached to the SEM.

3. RESULTS

3.1. Open Circuit Potential (OCP)

Figures 2-4 summarize the potential measurements (OCP) for both base and weld metal specimens after immersion for 12 weeks in both working conditions (stressed and unstressed samples) as a function of sulfide ions content. It is concluded that the OCP values for stressed samples were more negative than the unstressed ones. The OCP values ranged from −680 mV up to −730 mV. There was a slight change in potential values by increasing the sulfide content, where the OCP values were in the range of ~ −680 to −700 mV. The most negative OCP was in case of NS-4 solution containing 0.4 mM Na_2S, as shown in Figure 4.

3.2. Weight Loss Measurements

Figure 5 illustrates a summary of corrosion rate (mpy) based on weight loss measurements for base as well as weld metal coupons after immersion for 3 month, as a function of sulfide content. For base metal coupons, the highest corrosion rate was obtained in NS-4 + 0.4 mM Na_2S

(0.09 mpy). Increasing the sulfide ions content above 0.4 mM, had an inhibition effect on the corrosion behavior of carbon steel. The lowest corrosion rate (0.06 mpy) was obtained in NS-4 + 4 mM Na_2S. Similarly, for weld metal coupons, the highest corrosion rate (0.18 mpy) was obtained in NS-4 + 0.4 mM Na_2S. Weld metal coupons showed higher corrosion rate than its correspondents of base metal immersed in the same solutions.

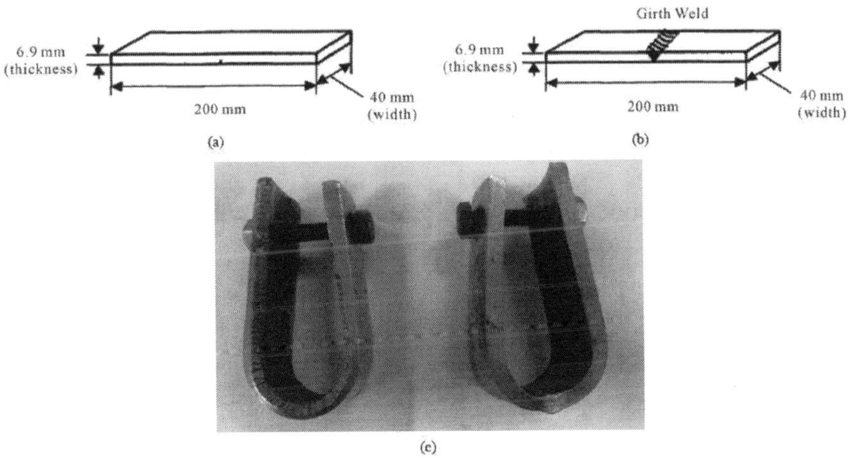

Figure 2. Test Specimens. (a) Base-metalpecimen; (b) Weld Specimen; (c) U-bend stressed specimens.

3.3. Potentiodynamic Test

The electrochemical parameters calculated using the Tafel plots (Figure 6) confirmed that the addition of sulfide increases the corrosion current, i_{corr} and consequently the corrosion rates (mpy). The anodic branches present a uniform dissolution for examined specimens in all solutions.

Corrosion data obtained from Tafel's plot are summarized in Table 5. From this table, the extent of increase in i_{corr} and corrosion rates (mpy) is found to be a function of the concentration of Na_2S. The addition of Na_2S has an inhibition effect of the corrosion behavior of carbon steel samples. This trend is in agreement with test result based on weight loss measurements. Also from Table 5, and in agreement with the OPC measurement, the most negative corrosion potential (−768 mV) was obtained in NS-4 solution containing 4 mM Na_2S.

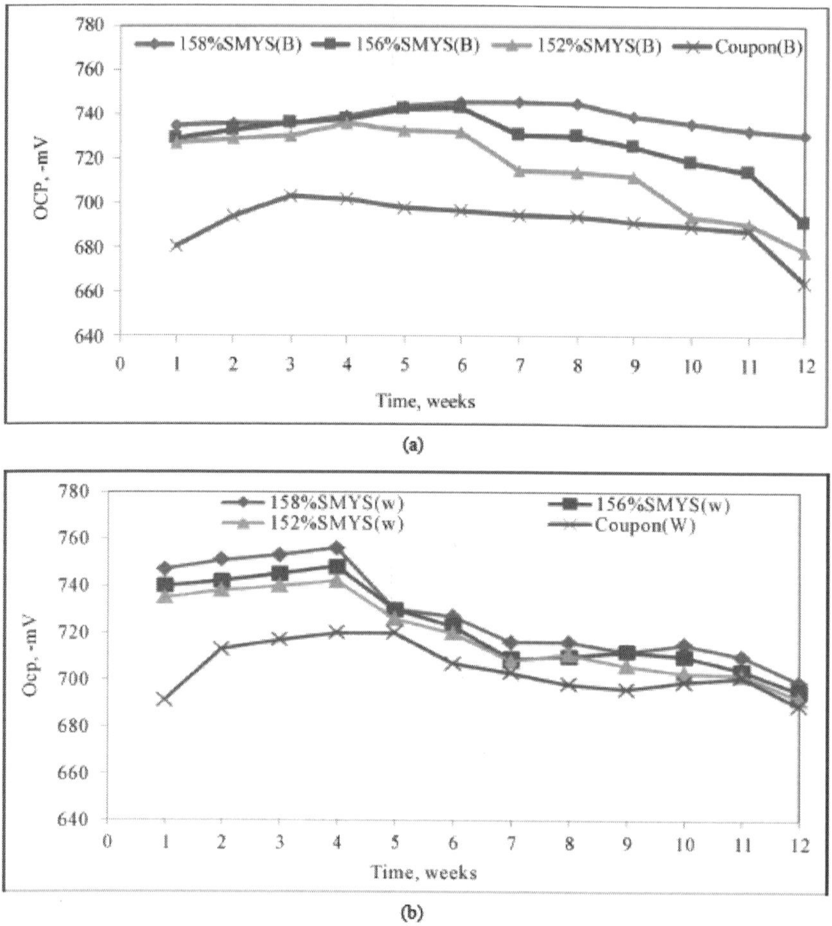

Figure 3. Effect of plastic deformation applied on OCP measurements in NS-4 solution. (a) Base-metal; (b) Welded-metal.

Corrosion rate (mpy) calculated from the Tafel's plots are presented in Figure 7. According to Figure 7, the weld metal specimen had higher corrosion rate than base metal one for both working conditions (stressed and unstressed). In case of stressed welded samples, the corrosion rates were 18, 32, and 15 mpy for NS-4 solutions with 0, 0.4, and 4 mM Na_2S, respectively.

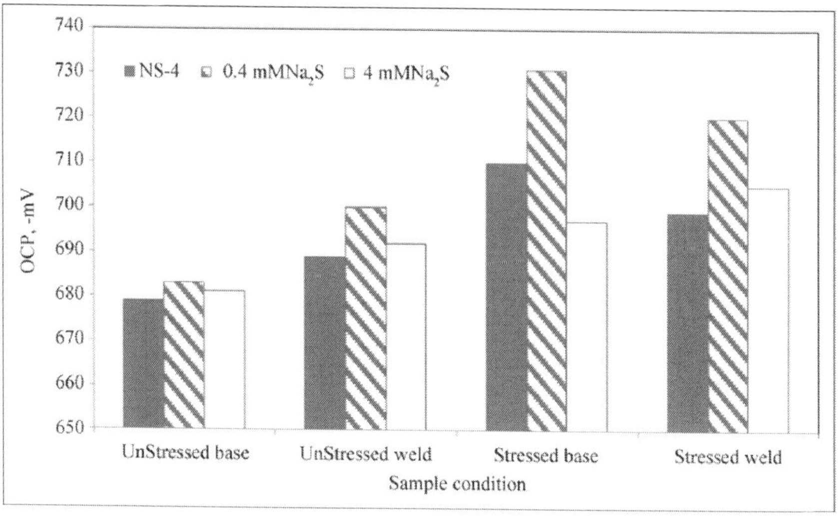

Figure 4. Open circuit potential of specimens for different sulfide-contaminated environments and different samples preparation conditions after immersion for 12 weeks.

3.4. SEM Surface Morphology and EDAX Analysis

The SEM images of base as well as weld metal coupons after immersion for three month in different sulfide polluted environments are illustrated in Figure 8. The SEM micrograph of base metal coupon which was immersed in NS4 solution, is characterized by dense layer (Figure 8(a)). From Table 6, EDAX analyses showed a high content of iron (88.14%) and the presence of oxygen (6.05%), carbon (4.58%), and manganese (%1.23). On the other hand, the SEM micrograph of weld metal coupon (Figure 8(b)) is characterized by flowery shaped layer formed on the metal surface immersed in the same above environment.

Figure 8(c) shows the SEM micrograph of weld metal coupon which was immersed in NS-4 + 0.4 mM Na_2S after 3 months. The SEM image shows a thin porous layer, and consequently non protective layer. EDAX analysis of this layer showed a lower content of iron (50.1%) and the presence of higher content of oxygen (32.7%), carbon (9.08%), magnesium (3.81%) and sulfur (4.17%), by increasing the sulfide content.

Figure 8(d) shows the SEM micrograph of the surface of the weld metal coupon which was immersed in NS-4 + 2 mM Na_2S for 3 months. The SEM image shows a dense spongy layer formed on the metal surface.

Figure 8(e) shows the SEM micrograph weld metal coupon which was immersed in NS-4 + 4 mM Na_2S for 3 months. The SEM image shows a dense layer with some bright particles formed on the metal surface. EDAX analysis (Table 6) indicated a lower iron content (41.47%) and the presence of of oxygen (27.9%), carbon (9.05%), magnesium (4.54%) and higher sulfur content (15.71%), in case of 4 mM Na_2S.

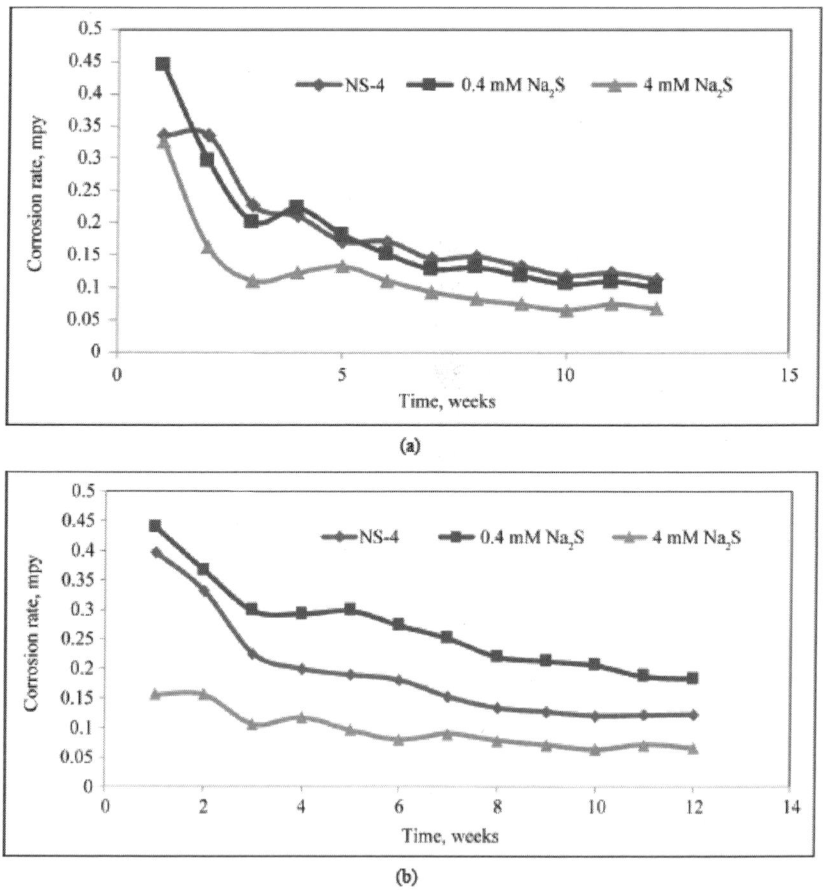

Figure 5. Corrosion rates (mpy) of welded-coupons after immersion in different sulfide- contaminated environments for 12 weeks-based on weight loss measurements. (a) Base metal; (b) Welded metal.

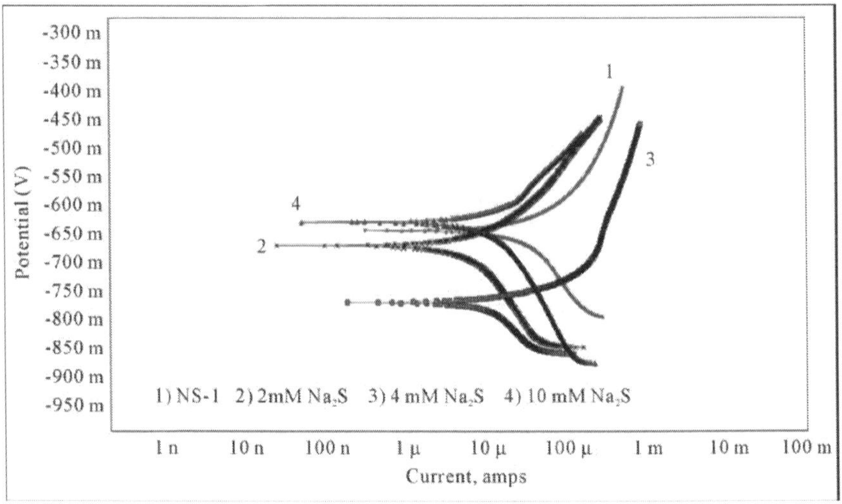

Figure 6. Polarization curves of weld coupons measured in sulfide contaminated environments.

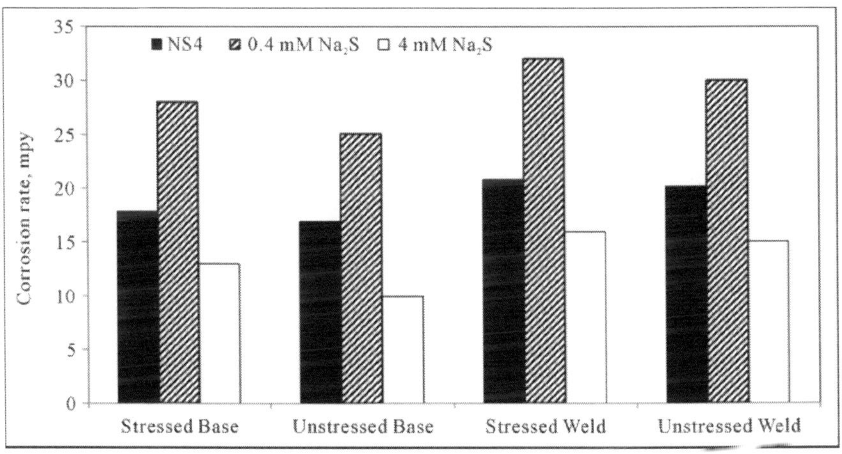

Figure 7. Corrosion rates of carbon steel specimens as a function of sulfide content as well as stresses applied.

Table 5. Corrosion data of API5LX-52 welded coupons immersed in different sulfide polluted electrolytes.

Condition	I corr ($A/\mu cm^2$)	Corrosion rate (mils/year)	Ecorr (mV)	βa mv/decade	βc mv/decade
NS4	21.2	20.2	−642	130	160
NS4 + 0.4 MNa_2S	31	30	−720	162	170
NS4 + 2 mNa_2S	15.34	14.74	−667	151.5	200
NS4 + 4 mNa_2S	15.5	15	−768	37.4	162.8
NS4 + 10 mNa_2S	18.8	18.1	−627	180	175

Figure 8(f) shows the SEM micrograph of the surface of the weld metal coupon which was immersed in NS-4 + 10 mM Na_2S for 3 months. The SEM image shows a dense protective layer with many bright particles formed on the metal surface.

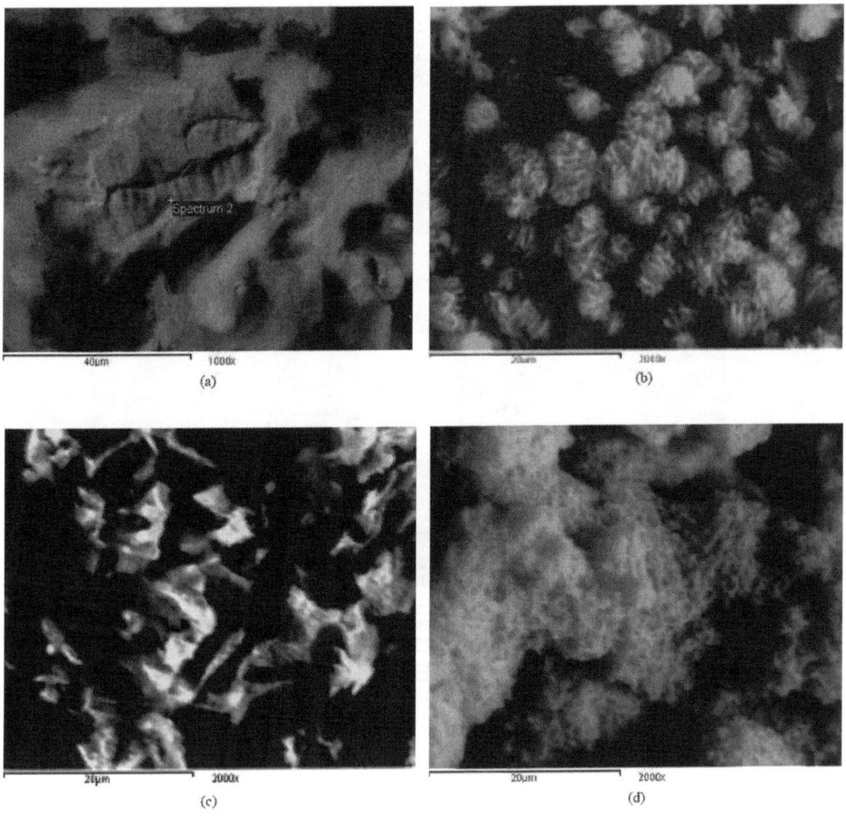

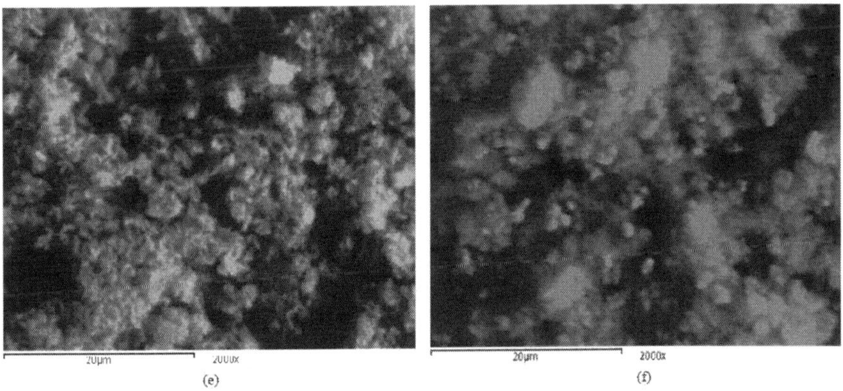

Figure 8. SEM of weld-metal coupons after immersion in sulfide contaminated environments for 3 month.

Table 6. EDAX analysis of API5LX-52 coupons taken at the weld zones after 3 month of immersion.

Condition	% C	% O	% Mn	% Mg	% S	% Si	% Fe
NS-4	4.58	6.05	1.23	0.0	0.0	0.0	88.14
NS-4 + 0.4mM Na_2S	9.08	32.79	0.0	3.81	4.17	0.0	50.15
NS-4 + 4 mM Na_2S	9.05	27.90	0.0	4.54	15.71	0.0	41.47

4. DISCUSSION

According to this work, and from potentiodynamic polarization test, it was noticed that stressed specimens showed higher corrosion rate than unstressed coupons in both cases of base and weld metal specimens. This is in agreement with other study showing that the corrosion current increased significantly and monotonically for samples with increasing plastic deformation [9] . The value of applied strain was based on the following formula;

$$\varepsilon = t/2R \text{ when } t < R \quad (2)$$

where ε is the applied strain, t is the specimen thickness and R is the radius of the bend [10] .

When steels are plastically deformed, some fraction of the deformation energy (~5%), is retained internally. The major portion of this stored energy is as strain energy associated with dislocations. The increase in the amount of strain energy would allow easier dissolution on specific sites since this energy is released on dissolution. This causes an increase in the corrosion rate with an increase in the amount of cold work [11] .

It was also recognized from potentiodynamic polarization test, and in agreement with weight loss test, that weld metal specimen suffered higher corrosion rate than base metal ones. This can be arisen from the fact that weld-metal composition (which is normally optimized for mechanical properties) tend to be slightly anodic to the parent steel in certain environments. Therefore, the weld metal corrodes at a higher rate than the base metal [12] [13] .

The deterioration of metal due to contact with hydrogen sulfide (H_2S) and moisture is called sour corrosion which is the most damaging to drill pipe. Although H_2S is not corrosive by itself, it becomes a severely corrosive agent in the presence of water, leading to pipeline embrittlement. Hydrogen sulfide when dissolved in water is a weak acid, and therefore, it is a source of hydrogen ions and becomes corrosive. The corrosion products are iron sulfides (FeS_x) and hydrogen. Iron sulfide forms a scale that at low temperature can act as a barrier to slow corrosion. The forms of sour corrosion are uniform, pitting, and stepwise cracking. The general equation of sour corrosion can be expressed as follows [14] :

$$H_2S + Fe + H_2O \rightarrow FeS_x + 2H + H_2O \qquad (3)$$

There is a possibility for formation of H_2S in the presence of some sulfide salts, according to the following equations [15] [16] ;

$$Na_2S \rightleftarrows Na^{+} + S^{-2} \qquad (4)$$

$$S^{-2} + H_2O \rightleftarrows HS^- + OH^- \tag{5}$$

$$HS^- + H_2O \rightleftarrows H_2S + H_2O \tag{6}$$

Another possible mechanism for iron dissolution in aqueous electrolytes containing H_2S is based on the formation of a tetragonal ferrous Sulfide (mackinawite) FeS as follows;

$$Fe + H_2S + H_2O \rightleftarrows FeHS - ads + H_3O^- \tag{7}$$

$$\text{FeHS-ads} \rightleftarrows FeHS^+ + 2e^- \tag{8}$$

$$FeHS^+ + H_3O^+ \rightleftarrows Fe + H_2S + H_2O \tag{9}$$

$$Fe^{+2} + HS^- \rightleftarrows FeS + H^+ \tag{10}$$

The FeS layer of steel is not stable, it is removed from the steel surface in an acidic environment, forming again H_2S, enhancing corrosion:

$$FeS + 2H \rightleftarrows Fe^{2+} + H_2S \tag{11}$$

There is a possibility that the surface of the steel will be covered by Fe_3S_4 protective layer that blocks the carbon steel from further corrosion. This layer could be doped with other corrosion products:

$$3FeS + H_2S\ ads \rightarrow Fe_3S_4 + 2H^+ + 2e^- \tag{12}$$

Subsequently

$$3Fe^{2+} + 3S^{-2} + H_2S \rightarrow Fe_3S_4 + 2H^+ + 2e^- \qquad (13)$$

This is in agreement with the study mentioned in the report from ICPTT [16] , and explains the reduction in corrosion rate measured by weight loss as well as Tafel's extrapolation technique, by increasing the sulfide ions content.

Increasing the sulfide ions content above 0.4 mM, had an inhibition effect on the corrosion behavior of carbon steel. For base metal coupons, the highest corrosion rate was obtained in NS-4 + 0.4 mM Na_2S (31 mpy). The lowest corrosion rate was obtained by conducting potentiodynamic test in NS4 + 4 mM Na_2S (~15 mpy).

The weight loss test of coupons immersed in NS4 + 4 mM Na_2S agreed with these results obtained from potentiodynamic test, which showed a low corrosion rate at high concentration of Na_2S. The reason for this low corrosion rate may be because of the use of sodium sulfide as the source of H_2S. At higher pH sodium sulfide may not produce H_2S gas and the dominant species could be S^{2-}, and the absence of H^+ ions decreases the corrosion rate. Also the reaction product, formed at higher pH is metal sulfides, which when precipitates on the surface of the metal gives protection to the metal as these metal precipitates have extremely low solubility. This is in agreement with a study performed by M. Koteeswaran [17] on CO_2 and H_2S corrosion in oil pipe lines.

In addition, some results demonstrated that the presence of hydrogen sulfide can inhibit the corrosion of iron under certain special conditions due to the formation of a ferrous sulfide—pyrite protective film [18] - [20] .

Recently, Sherar et al. [20] investigated the effect of inorganic sulfide on carbon steel corrosion in a solution containing chloride, bicarbonate and sulfate (pH 8.9) by following the evaluation of corrosion potential (Ecorr) and periodically measuring the polarization resistance (R_P) over an exposure period of a few months. When freshly-polished carbon steel was exposed directly to sulfide a low corrosion rate was observed. However, when sulfide was added to pre-corroded steel, the corrosion rate tripled [21] . These observations were consistent with the results reported by Newman et al. [21] , who hypothesized that a pre-corroded surface prevents FeS passi- vation.

The SEM image of the weld coupon immersed in NS-4 + 0.4 mM Na_2S showed a thin porous layer which is not protective formed on the steel's surface, and this is in support of high corrosion rate in this solution. The SEM image of the weld coupon at high concentrations of Na_2S showed a compact passive corrosion product film formed on the steel surface. The EDAX analysis showed that in case of NS-4 electrolyte, corrosion product formed on the metal surface was composed of; 4.58% C, 6.05% O, 1.23% Mn, 88.14% Fe. The protective product formed on the metal surface in case of 4 mM Na_2S is composed of; 9.08% C, 32.79% O, 3.81% Mg, 4.17% S, 50.15% Fe. It was also noticed that with the increase of sulfide concentration from 0.4 to 4 mM, the Fe/S ratio decreased from 12.02 at 0.4 mM, to 2.63 at 4 mM Na_2S. This shows that increasing the sulfide concentration leads to the formation of sulfur-rich iron sulfide [5] .

5. CONCLUSIONS

1) Increasing the sulfide ions content above 0.4 mM Na_2S, had an inhibition effect on the corrosion behavior of carbon steel, due to formation of protective film.

2) There was a slight change in potential values by increasing the sulfide content.

3) The protective product formed on the metal surface in case of NS-4 + 4 mM Na_2S was composed of; 9.08% C, 32.79% O, 3.81% Mg, 4.17% S, 50.15% Fe.

4) Increasing of sulfide concentration from 0.4 to 4 mM, the Fe/S ratio decreased from 12.02 at 0.4 mM to 2.6 at 4 mM. This showed that increasing the sulfide concentration led to the formation of sulfur-rich iron sulfide.

5) The open circuit potential values for stressed steels were more negative than the unstressed samples.

6) Weldments of carbon steel showed more tendency for corrosion than base steel.

ACKNOWLEDGEMENTS

This paper is extracted from a thesis submitted to the faculty of Engineering-Cairo University as partial fulfillment of the M.Sc. degree by one of the authors in 2015 (Eng. Mohamed Ali Farag).

REFERENCES

1 Mendili, Y.E., Abdelouas, A. and Bardeau, J. (2014) The Corrosion Behavior of Carbon Steel in Sulfide Aqueous Media at 30 ?C, JMEPEG, 23, 1350-1357.

2 Hamdy, A., Shoeib, M.A, Saehand, A.G. and Barakat, Y. (2008) The Electrochemical Behavior of Mild Steel in Sulfide Polluted NaCl at Different Velocities. International Journal of Electrochemical Science, 3, 1142-1148.

3 de Freitas, V., Lins, C., Leonard, M., Ferreira, M. and Alves Saliba, P. (2012) Corrosion Resistance of API X52 Carbon Steel in Soil Environment. Journal of Materials Research and Technology, 1, 161-166.

4 Ashour, E.A., Khorshed, L.A., Youssef, G.I., Zakria, H.M. and Khalifa, T.A. (2014) Electrochemical and Stress Corrosion Cracking Behavior of Alpha-Al Bronze in Sulfide-Polluted Salt Water: Effect of Environmentally-Friendly Additives. Materials Sciences and Applications, 5, 10-19.

5 Xuejun, X., Wanqin, Y., Shunan, C., Ling, P., Xunjie, G. and Keru, P. (2005) The Effect of Sulfide on the Corrosion Behavior of Carbon Steel in Natural Waters. Anti-Corrosion Methods and Materials, 52, 145-147.

6 Fouda, A.S., El-Khateeb, A.T., Ibrahim, M. and Fakih, M. (2015) Adhatoda Aqueous Plant Extract as Environmentally Being Corrosion Inhibitor for Carbon Steel in Sanitation Water in Polluted NaCl Solutions and Its Biological Effect on Bacteria. Nature and Science, 13, 71-82.

7 Parkins, R.N., Blanchard, W.K.J. and Delanty, B.S. (1994) Transgranular SCC of High Pressure Pipelines near Neutral pH. Corrosion, 50, 394-408.

8 Ovri, J.E.O., Okeahailam, S.I. and Onyemaobi, O.O. (2013) Microbial Corrosion of Mild and Medium Carbon Steel. Journal of Engineering Science and Technology, 8, 639-653.

9 Li, X.C., Eadie, R.L. and Luo, J.L. (2008) Influence of Plasticity on Corrosion and Stress Corrosion Cracking Behavior in near Neutral pH Environment. 43, 297-303.

10 (2009) ASTMG30, Standard Practice for Making and Using U-Bend Stress-Corrosion Test Specimens.

11 Hertzberg, R.W. (1996) Deformation and Fracture Mechanics of Engineering Materials. 4th Edition, John Wiley & Sons, Inc., Hoboken.

12 Lee, C.-M., Bond, S. and Woollin, P. (2005) Preferential Weld Corrosion, Effects of Weldment Microstructure and Composition. Proceedings of Corrosion 2005, NACE International, Paper 05277.

13 Popoola, L.T., Grema, A.S., Latinwo, G.K., Gutti, B. and Balogun, A.S. (2013) Corrosion Problems During Oil and Gas Production and Its Mitigation. International Journal of Industrial Chemistry, 4, 1-15.

14 Yuan, G., Wang, S., Zhang, Y. and Huang, Y. (2009) Role of Sulfide Ion on Corrosion Performance of Pipeline Steel in Simulated Inner-Seeper Environment. International Conference on Pipelines and Trenchless Technology ICPTT 2009, Shanghai, 18-21 October 2009, 1643-1651.

15 Tang, J., Shao, Y., Zhang, T., Meng, G. and Wang, F. (2011) Corrosion Behavior of Carbon Steel in Different Concentrations of

HCI Solutions Containing H2S at 90 ?C. Corrosion Science, 53, 1715-1723.

16 Koteeswaran, M. (2010) CO2 and H2S Corrosion in Oil Pipelines. M.Sc. Thesis, Faculty of Mathematics and Natural Science, University of Stavnger, Norway.

17 Arenas-Marteniz, L.F. (2012) Hydrogen Sulfide Corrosion of Weld Regions in API X52 Steel. Ingeniria, Investigacion y Tecnologia, 13, 473-478.

18 Ma, H., Cheng, X., Chen, S., Wang, C., Zhang, J. and Yang, H. (1998) An AC Impedance Study of the Anodic Dissolution of Iron in Sulfuric Acid Solutions Containing Hydrogen Sulfide. Journal of Electroanalytical Chemistry, 451, 11-17.

19 Ma, H., Cheng, X., Li, G., Chen, S., Quan, Z., Zhao, S. and Niu, L. (2000) The Influence of Hydrogen Sulfide on Corrosion of Iron Under Different Conditions. Corrosion Science, 42, 1669-1683.

20 Sherar, B.W.A., Keech, P.G., Noel, J.J. and Shoesmith, D.W. (2013) The Effect of Sulphide on Carbon Steel Corrosion Behavior in Anaerobic Near-Neutral Saline Solutions. Corrosion, 1, 67-76.

21 Newman, R.C., Rumash, K. and Webster, B.J. (1992) The Effect of Pre-Corrosion on the Corrosion Rate of Steel in Neutral Solutions Containing Sulfide: Relevance to Microbially Influenced Corrosion. Corrosion Science, 33, 1877-1884.

Index